Johan Vindevogel · Pat Sandra

Introduction to Micellar Electrokinetic Chromatography

Chromatographic Methods

Series editors

W. Bertsch

H. Frank

W.G. Jennings

P. Sandra

Introduction to Micellar Electrokinetic Chromatography

By
Johan Vindevogel and Pat Sandra

Hüthig Buch Verlag Heidelberg

Johan Vindevogel
Pat Sandra
Laboratory of Organic Chemistry
State University Gent
Krijgslaan 281, S4
B-9000 Gent
Belgium

Die Deutsche Bibliothek - CIP-Einheitsaufnahme

Vindevogel, Johan
Introduction to micellar electrokinetic chromatography / by
Johan Vindevogel and Pat Sandra. - Heidelberg : Hüthig, 1992
(Chromatographic methods)
ISBN 3-7785-0861-X
NE: Sandra, Pat:

Preface

For a long time, electrophoresis and chromatography have been living separate lives, as if each were confined to its own niche. With the advent of miniaturization the two techniques have again found common ground. This process of convergence has been further promoted by explicitly combining chromatographic selectivity with electrokinetic propulsion. Micellar Electrokinetic Chromatography has so far achieved a remarkable degree of success. However, an introduction to this new technique should not be construed as a proclamation of its maturity – indeed many new developments are to be expected. And yet there already exists a considerable body of literature that can be made more accessible by collecting it into an organized entity.

We wish to express our thanks to our colleagues and the technical staff of our laboratory who all contributed, in one way or another, to this work. Special thanks go to Roman Szücs (Czechoslovakia) and Francisco Rojos Callejas (Mexico) for useful discussions and their help in performing experiments. We are also indebted to the many authors and publishers who have consented to our using figures from published material.

We thank the Belgian *National Fund for Scientific Research* and the *National Lottery* for financial support to our laboratory.

Johan Vindevogel Gent, January 1992
Pat Sandra

Introduction

More than ten years have now elapsed since the possibility of Capillary Zone Electrophoresis (CZE) was demonstrated. Six years ago, a variant on this technique, Micellar Electrokinetic Chromatography (MEKC), was developed, bridging the gap between electrophoresis and chromatography. In both cases, the rather quiet development during the first two or three years was followed by an outburst of activity within the chromatographic community and within in a short time a considerable amount of knowledge and experience has been collected. The recognition of, and the faith in the potential value of, those new separation techniques are exemplified by the rapid development of commercial instrumentation, in sharp contrast to the lack of a similar evolution in miniaturized liquid chromatography.

This book is primarily devoted to Micellar Electrokinetic Chromatography. As it is intended for those who have a background in chromatography but not in electrophoresis, the basic principles of electrophoresis in general but also, more specifically, of Capillary Zone Electrophoresis are included. As such, it can also be used as a short introduction to the latter, although topics like protein analysis and Capillary Gel Electrophoresis are not included.

The material in this book is organized into four parts. The *first* part deals with fundamental aspects of electrokinetic analysis : analytical formats (Chapter 1) and the principles governing electrophoretic and electroosmotic mobility (Chapter 2). The application of those principles to obtain separation is described for different techniques : Capillary Zone Electrophoresis (Chapter 3), Capillary Electro-chromatography (Chapter 4) and Micellar Electrokinetic Chromatography (Chapter 5). The *second* part focuses on instrumental aspects, covering thermal effects (Chapter 6), columns (Chapter 7), injection techniques (Chapter 8) and detection methods (Chapter 9). The *third* part describes the resolution optimization in Micellar Electrokinetic Chromatography. Of the three traditional parameters : capacity factor, selectivity and efficiency only the latter is treated in a separate chapter (Chapter 10). Given the pecularities of resolution optimization in Micellar Electrokinetic Chromatography, the role of the capacity factor and of selectivity are discussed together (Chapter 11). Finally, a review of published applications has been collected in the *fourth* part : one chapter is devoted to general applications (Chapter 12) while the other deals specifically with chiral separations (Chapter 13). This part is not intended as a manual of analysis but serves to demonstrate the possibilities of

Micellar Electrokinetic Chromatography. Therefore, the sections in it cover groups of compounds that are chemically or functionally related, whatever was considered most appropriate.

1 General Principles of Electrokinetic Analysis

1.1 The Miniaturization of Electrophoresis

Electrophoresis [1-4] is a separation technique based on the differences in ion linear velocity v in an electric field E. The ion linear velocity is determined by the intrinsic ion mobility μ and the applied voltage V (over a length L) :

$$v = \mu E = \mu \frac{V}{L} \tag{1.1}$$

Through the movement of ions (electric current) frictional energy is released (Joule effect) by which the solution warms up. As the system will loose part of that heat at its boundaries, thermal gradients will be developed and, in turn, convection will be generated which can lead to remixing of the separated zones. The classical[1] solution to this fundamental problem in electrophoresis is the use of stabilizing media. Instead of performing electrophoresis in free solution, a porous matrix is used which hinders convective motion : gel electrophoresis and paper electrophoresis.

Gel electrophoresis, developed around 1950 by *Tiselius* [5] and generally using polyacrylamide or agarose gels, has grown into a powerful analytical technique with high resolving capability. Gel electrophoresis is now the method of choice for the analysis of proteins and nucleic acids. Nevertheless, some disadvantages remain. Gels must be carefully prepared to control their pore size characteristics and can be used only once. Analysis is generally time consuming and the results cannot be monitored on-line but require subsequent staining and scanning steps.

Alternatively, the separation compartment can be miniaturized, facilitating heat exchange through the increased surface-to-volume ratio and reducing the extent of the thermal gradients. Some key steps in this approach, which has led to a convergence between electrophoretic and chromatographic practice, are summarized in **Figure 1.1**.

In 1967, *Hjertén* [6] counteracted gravity driven convection by rotating a 3 mm ID glass tube, allowing electrophoretic separations without using stabilizing gels.

---

1 'Classical ' will be continuously used here to refer to non-capillary forms of electrophoresis, however without implying or suggesting that this field would be no longer subject to innovations.

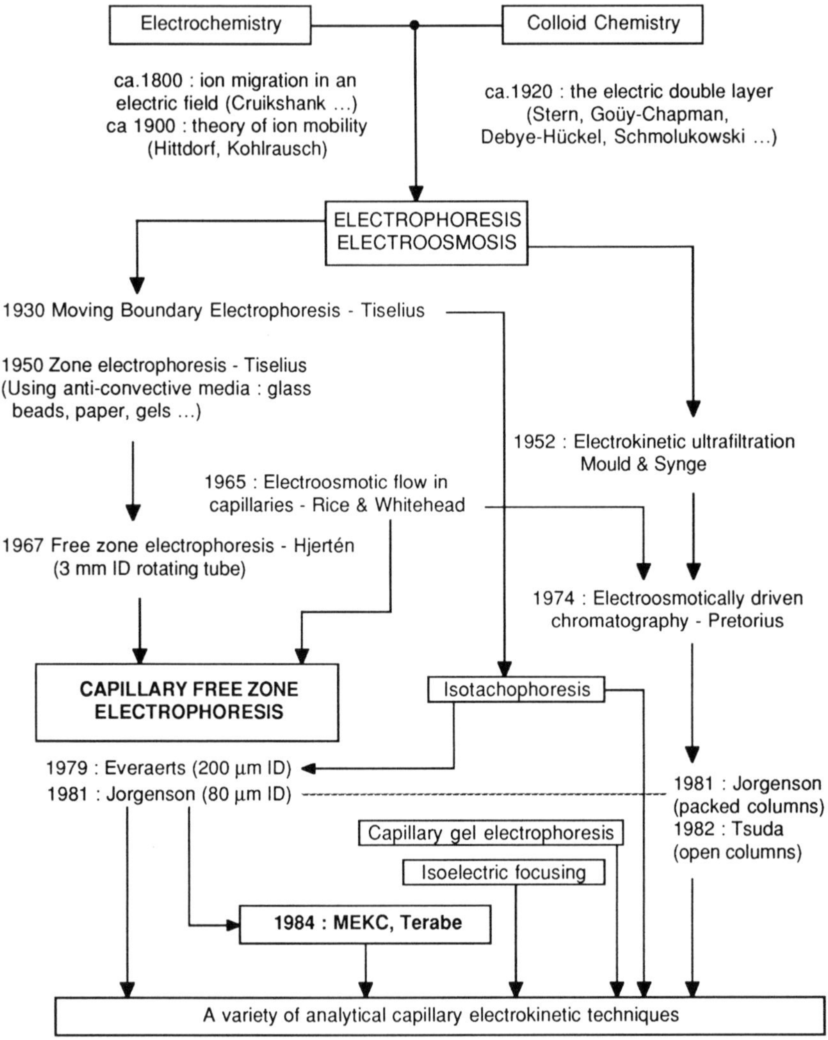

Figure 1.1
Some key steps in the development of capillary electrophoretic techniques.

He also developed a scanning device for the UV detection of the zones as they moved through the tube. From the side of chromatography, *Pretorius* recognized as long ago as 1974 the potential of electroosmosis, the bulk flow that occurs when a contained fluid is subjected to an electric field, as a propulsive principle in elution chromatography [7]. At the same time he remarked that the inappropriateness of injection and detection systems at that time prevented its full utilization. Instrumental miniaturization was clearly a key problem in designing electrodriven systems that could effectively dissipate Joule heat.

In 1979, exploiting the degree of miniaturization that had already taken place in isotachophoresis, *Everaerts* demonstrated the possibility of free zone electrophoretic separations in 200 μm ID Teflon columns with on-line detection [8]. In 1981, almost parallel with the onset of the miniaturization trend in liquid chromatography [9], highly efficient protein separations were realized by *Jorgenson*, using 80 μm ID glass columns [10]. Since then, capillary electrokinetic techniques have rapidly evolved, allowing highly efficient separations in many different formats.

1.2 Formats in Capillary Electrophoretic Analysis

Mobility differences can be exploited in several ways to obtain separation. Using the same basic instrumentation (**Figure 1.2**), specific analytical conditions have resulted in different methods. But those methodologies can equally well be considered as processes. It has been remarked that in one method, at particular stages or under particular circumstances, processes occur that are typical for another method [11].

The following enumeration is limited to those formats that have found application in capillary techniques. Among the ones mentioned, Zone Electrophoresis and Micellar Electrokinetic Chromatography will be discussed in depth in subsequent chapters.

1.2.1 Zone Electrophoresis

In zone electrophoresis, a carrier or background electrolyte is used to maintain a constant electric field along the separation axis (**Figure 1.3**). Due to this constant electric field, sample ions will migrate with a linear velocity proportional to their mobility. When mobilities differ, separate zones are formed. The carrier electrolyte prevents the formation of non-conductive gaps in between the zones.

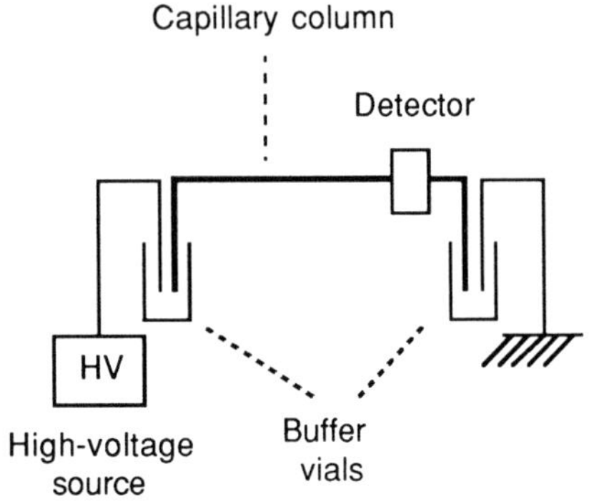

Figure 1.2
Basic lay-out for a capillary electrophoresis instrument.

The zones are in fact similar to what are known to chromatographers as peaks : they represent the analyte distribution (caused by zone or peak spreading) around a sharply defined 'ideal' analyte position. Preferably, the ion concentration in the sample is much lower than that of the background, so that the local conductivity, and the electrical field, is not disturbed by the presence of the zone. Under these conditions, all the zones move at a constant, characteristic speed.

When the column is filled with a buffer, as was the case in the experiments of *Hjertén*, *Everaerts* and *Jorgenson*, the technique is known as Free Zone Electrophoresis. As already remarked, this is rarely applied in classical electrophoresis, where the use of a stabilizing medium is preferred. In that case, the restrictive gel influences actual mobilities according to molecular weight, shifting the separation mechanism to size fractionation. In the case of proteins, this is further emphasized by denaturation with charged surfactants, mostly sodium dodecyl sulfate (SDS/PAGE or SDS-polyacrylamide gel electrophoresis [1]), resulting in a constant charge-to-mass ratio. This separation mechanism has been exploited in Capillary Gel Electrophoresis using gel-filled capillaries [12].

Several attempts have been made to combine electrophoretic and chromatographic separation modes. Most successful until now has been Micellar Electrokinetic Chromatography [13]. Although an additional selectivity principle is involved, the basic electrophoretic processes are those of zone electrophoresis. The only difference is that a charged surfactant is added to the buffer. The resulting micelles act as a pseudo-stationary phase, interacting by hydrophobic association with neutral analyte molecules, carried along with the electroosmotic flow. In this way, neutral molecules can be separated in an electric field.

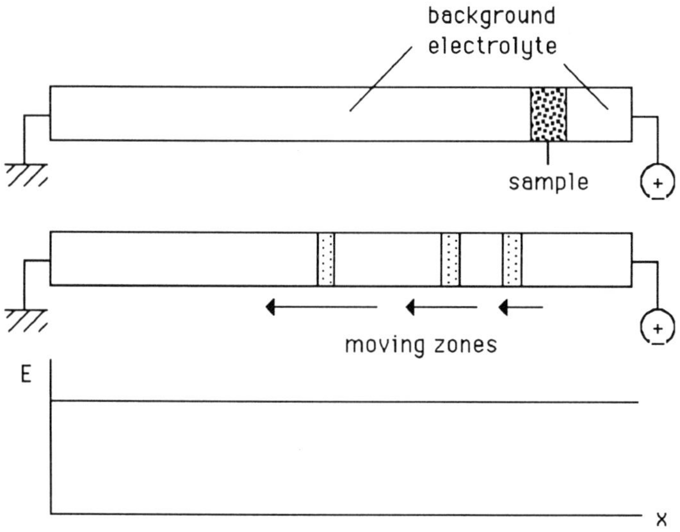

Figure 1.3
Zone Electrophoresis. Separation of sample mixture into zones and value of field strength E along the separation axis.

1.2.2 Isotachophoresis

In this method [14], the sample mixture is applied between a leading and a terminating electrolyte (**Figure 1.4**). Assuming the analysis of cations (and for simplicity, that only a single kind of counter-ion is present in the system), the leading cation is selected to be more mobile than any sample cation, and correspondingly, the terminating cation is the least mobile. As mobility, and consequently conductivity or resistance, varies along the separation axis, the field strength is not homogeneous. In zone electrophoresis the separation chamber can be represented by a single resistor, but in isotachophoresis it must be considered as a set of resistors, connected in series, each with a characteristic voltage drop.

Initially, the components in the sample mixture will move with their own characteristic mobility. During that stage, the leading cations move faster than the sample ions. In between the leading ions and the sample ions local resistance and

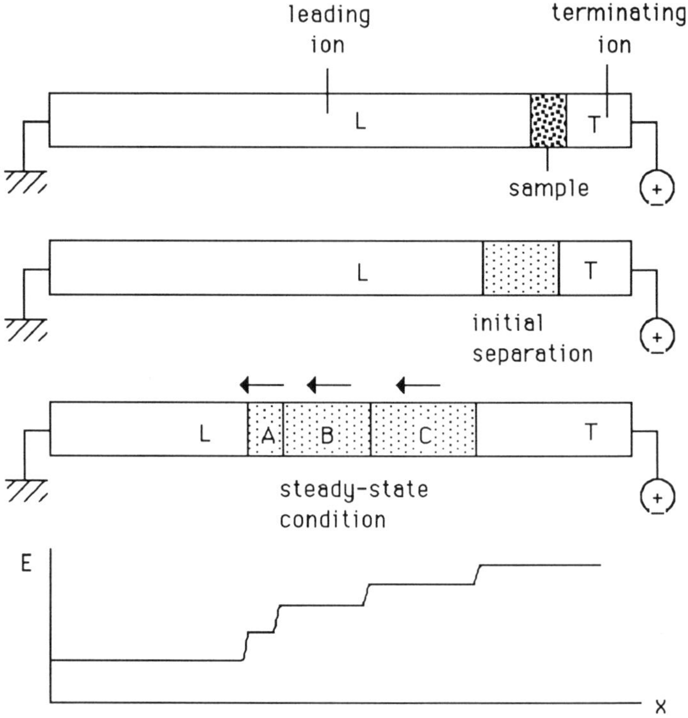

Figure 1.4
Isotachophoresis. Application of sample mixture, initial separation stage, final steady-state condition and value of field strength E along the separation axis.

electric field will increase as there is no carrier electrolyte (and the solvent is essentially non-conductive). This speeds up the faster moving sample ions and they are pulled along to avoid the formation of a non-conductive gap. In turn, they act as a leading electrolyte for the slower moving sample ions. Eventually, a steady state condition is obtained in which each ion occupies its own zone, adjacent to a zone with higher mobility on one side and one with lower mobility on the other side. But the actual velocity of all the zones is equal (iso-tacho-phoresis = same + speed + movement). The whole process is governed by Ohm's law and requirements of electroneutrality, as summarized by the Kohlrausch regulating function, and necessitates the adjustment of the ion concentration in each zone. Signal processing differs from chromatography. The equilibrium concentration in a zone, or signal intensity, is a qualitative criterion, while the zone length is a quantitative criterion.

This very simplified description does not render justice to the advantages of isotachophoresis : self-sharpening of the zone boundaries (making the injection procedure and thermal effects less critical) and concentration effects (allowing large, but dilute, samples). Isotachophoresis is also referred to as displacement electrophoresis and the analogy with displacement chromatography [15] goes further than that. Neither technique follows a much regretted rule in elution chromatography that separation goes along with dilution. Notwithstanding this advantage, both techniques have failed to become widespread beyond specialized laboratories as their successful application is dependent on the judicious control of separation conditions, often requiring much information on the sample. In the case of isotachophoresis, rational selection of leading and terminating electrolytes depends on the knowledge of mobilities and pK_a values for all sample components.

Even if not used as a *method*, some understanding of the *process* is necessary. Whenever conductivity discontinuities occur in zone electrophoresis (e.g. when sample and run buffer are different), isotachophoretic effects must be considered. Therefore a simplified description of the Kohlrausch regulating function, borrowed from [14], has been included in Appendix 1.

1.2.3 Isoelectric Focusing

The use of isoelectric focusing [16] is limited to the separation of amphoteric molecules like proteins. An ampholyte mixture is used as the background buffer. It consists of amphoteric molecules (often polyamino-polycarboxylic acids) with a range of pI-values, covering the desired pH interval. Along the separation axis, a pH gradient is formed. Sample constituents will migrate, constantly adapting their charge to the local pH (**Figure 1.5**). When the local pH is equal to the isoelectric point, the protein will no longer migrate (as its charge has become zero) and remain focused in a narrow band.

When isoelectric focusing is performed in a gel, as is the case in classical electrophoresis, the focused proteins can be visualized by staining. When performed in a carrier-free capillary [17], an additional step is necessary to force the zones to pass at the detector. This can be accomplished by inducing hydrodynamic flow or electrophoretically, by changing the anolyte or catholyte buffers, breaking down the pH gradient [17,18].

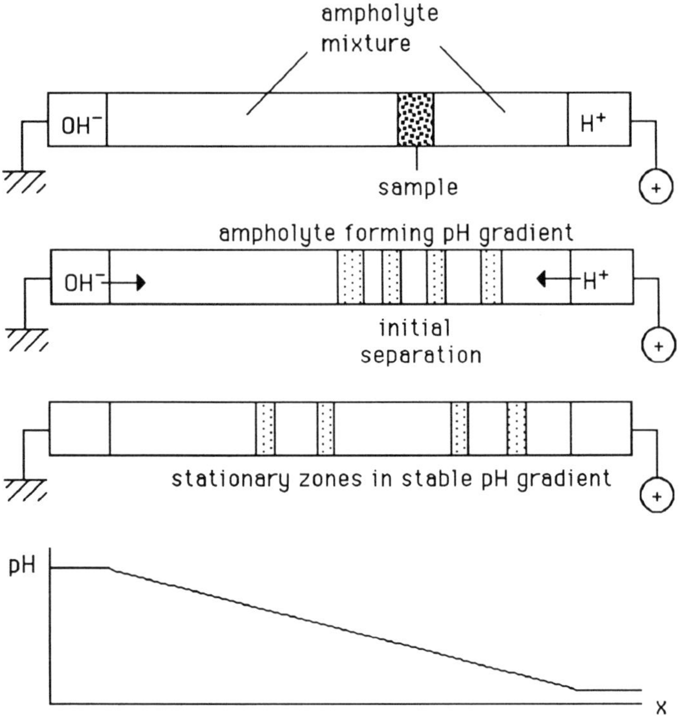

Figure 1.5
Isoelectric focusing. Application of sample mixture, initial separation and formation of the pH gradient, final stationary zone condition and value of pH along the separation axis.

References

[1] A.T.Andrews, "Electrophoresis : Theory, Techniques, and Biochemical and Clinical Applications", Oxford Science Publications, 2nd ed., 1982.
[2] S.Hjertén, Topics Bioelectrochem. Bioenerg., **2** (1978), 89.
Analytical electrophoresis.
[3] J.W.Jorgenson, Anal.Chem., **58** (1986) 743A.
Electrophoresis.
[4] O.Vesterberg, J.Chromatogr., **480** (1989) 3.
History of electrophoretic methods.
[5] S.Hjertén, Chromatogr.Rev., **9** (1967) 122.
Free zone electrophoresis.
[6] S.Hjertén, Electrophoresis (Weinheim), **9** (1988) 3.
The history of the development of electrophoresis in Uppsala.
[7] V.Pretorius, B.J.Hopkins, J.D.Schieke, J.Chromatogr., **99** (1974) 23.
Electro-osmosis. A new concept for high-speed liquid chromatography.
[8] F.E.P.Mikkers, F.M.Everaerts, Th.P.E.M.Verheggen, J.Chromatogr., **169** (1979) 11.
High-performance zone electrophoresis.
[9] M.V.Novotny, D.Ishii (eds), "Microcolumn separations", Elsevier Science Publ., Amsterdam, 1985 (J.Chromatogr.Lib., vol.30).
[10] J.W.Jorgenson, K.D.Lukacs, J.High Resolut.Chromatogr.Chromatogr. Commun., **4** (1981) 230.
Zone electrophoresis in open-tubular glass capillaries : Preliminary data on performance.
[11] J.L.Beckers, F.M.Everaerts, J.Chromatogr., **508** (1990) 19.
Isotachophoresis with two leading ions and migration behaviour in capillary zone electrophoresis. II. Migration behaviour in capillary zone electrophoresis.
[12] A.S.Cohen, B.L.Karger, J.Chromatogr., **397** (1987) 409.
High-performance sodium dodecyl sulfate polyacrylamide gel capillary electrophoresis of peptide and proteins.
[13] S.Terabe, K.Otsuka, K.Ichikawa, A.Tsuchiya, T.Ando, Anal.Chem., **56** (1984) 111.
Electrokinetic separations with micellar solutions and open-tubular capillaries.
[14] F.M.Everaerts, J.L.Beckers, Th.P.E.M.Verheggen, "Isotachophoresis : Theory, Instrumentation and Applications", Elsevier Science Publ., Amsterdam, 1976 (J.Chromatogr.Lib., vol.6).
[15] C.Horváth, A.Nahum, J.H.Frenz, J.Chromatogr., **218** (1981) 365.
High-performance displacement chromatography.
[16] P.G.Righetti, E.Gianazza, C.Gelfi, M.Chiari, P.K.Sinha, Anal.Chem., **61** (1989) 1602.
Isoelectric focusing in immobilized pH gradients.
[17] S.Hjertén, M.-D.Zhu, J.Chromatogr., **346** (1985) 265.
Adaptation of the equipment for high-performance electrophoresis to isoelectric focusing.
[18] S.Hjertén, J.Liao, K.Yao, J.Chromatogr., **387** (1987) 127.
Theoretical and experimental study of high-performance electrophoretic mobilization of isoelectrically focused protein zones.

2 Factors Affecting Mobility

Separation in electrophoresis relies on mobility differences and the means to control them. Provided that the components of a mixture are known, knowledge of the mobilities could facilitate successful separation. Unfortunately, and in spite of the theoretical achievements, a priori estimation of mobilities is only possible for ions with a relatively simple geometry, like metal cations. An alternative approach, deriving mobility from conductivity data [1], is useful in isotachophoresis but fails to take into account the influence of the carrier electrolyte used in zone electrophoresis. Anyway, necessary physical data would be hard to find for most organic chemicals, if available at all. Therefore the subsequent discussion focuses on those aspects that allow evaluation of trends in mobility changes, ignoring the details of theories that would be hard to apply to more complex, organic ions.

2.1 Ion Mobility

2.1.1 The Electric Double Layer

The behaviour of a particle in an electric field is determined by its electrophoretic mobility μ, which is the linear velocity v per unit of applied field E :

$$\mu\left(\frac{m^2}{V.s}\right)=\frac{v\ (m/sec)}{E\ \ (V/m)} \tag{2.1}$$

The simplest description of ion mobility can be obtained by balancing electromotive and frictional forces (using Stokes' law for the latter) for an ion with charge q and radius r :

$$q\,E = 6\,\pi\,r\,\eta\,v \tag{2.2}$$

or

$$\mu = \frac{v}{E} = \frac{q}{6\,\pi\,r\,\eta} \tag{2.3}$$

Although this equation is useful to describe changes in mobility as a function of charge or molecular mass, it shows some limitations. Through Stokes' law, eq.(2.3) is only valid for rigid, spherical ions. For more complex, irregularly shaped ions the corresponding friction factors are not easily evaluated. Also, eq.(2.3) ignores other forces [1] originating in the counter-ion atmosphere, leaving the influence of the medium on the mobility unexplained. Therefore it is more accurate to describe the mobility as a function of the ζ (zeta) potential :

$$\mu = \frac{\varepsilon \,.\, \zeta}{6 \pi \eta} = \frac{2}{3} \frac{\varepsilon_0 \,.\varepsilon_r \,.\, \zeta}{\eta} \tag{2.4}$$

with :

ζ = zeta potential (volt)

ε = permittivity of buffer (= $4\pi\, \varepsilon_0.\varepsilon_r$),

ε_0 = permittivity of vacuum ($8.85\ 10^{-12}\ C^2/N.m^2$)

ε_r = relative permittivity or dielectric constant (± 78 for water at 25°C)

Unfortunately, the ζ potential of complex ions is not an easily accessible quantity and, like eq.(2.3), eq.(2.4) is of little use for mobility prediction. Instead, mobilities are measured to calculate ζ potentials (e.g. [2]). But at least, the physical meaning of the ζ potential allows qualitative evaluation of the influence of several parameters on ion mobility.

The ζ potential is a quantity derived from the double-layer theory [3-5], describing the interface between a single ion and its counter ions and the surrounding solution. Several theories have been established to describe the potential across the double layer, differing in the alleged structure of the interface. In the *Helmholtz* model (1879) the counter ions are considered to be closely associated with the central ion, a structure described as a capacitor with a linear decay of the potential. *Goüy* (1910) and *Chapman* (1913) however assumed freely moving counter ions resulting in an exponentially decaying potential. A similar model was used by *Debye* and *Hückel* (1923) to predict electrolyte activity coefficients, well known from physical chemistry. *Stern* (1924) combined the Helmholtz and Goüy-Chapman models, considering some of the counter ions to be fixed while the remaining form a diffuse layer (**Figure 2.1**). Under the influence of an electric field, the fixed ions and molecules will move along with the central ion, forming a shear plane with the surrounding medium. The ζ-potential is the value of the ion potential at the shear plane.

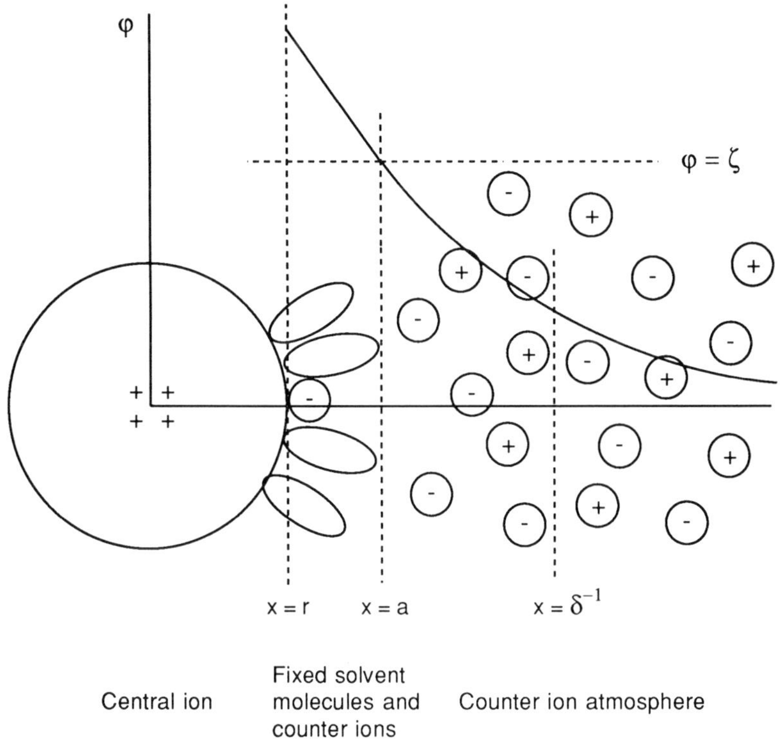

Figure 2.1
The Stern model (r = geometric ion radius, a = shear plane, δ^{-1} = Debye length).

Using a simple model, and ignoring the subtle differences between the treatments by Debye-Hückel, Goüy-Chapman or Stern, the qualitative effect of several parameters on ion mobility can now be described. Let the potential φ_1 and the associated ζ potential ζ_1 represent a reference situation (**Figure 2.2**). Increasing the size of the ion pushes the shear plane outward, reducing the ζ potential to ζ_2. For similarly charged ions, mobility will decrease with molecular weight. Increasing the charge of the ion increases its potential to φ_3 and the ζ potential to ζ_3, resulting in a mobility increase. Although those simple rules of thumb are in accordance with eq.(2.3), they should be used with care. Mobility data for simple inorganic ions (**Table 2.1**) show that mobilities are related to atomic weight (or crystal ionic radius) in the opposite way. Of course, differences in solvation (strongest for the smallest ion), and the consequence for the effective ionic radius (according to eq.(2.3)) or the position of the shear plane (according to the ζ potential model), must be taken into account. Fortunately, solvation is less important for organic ions, allowing estimation of the relative mobilities within homologous series. As an example, **Table 2.2** shows the mobilities of some alkyl ammonium ions. As predicted, mobility decreases with increasing substitution of hydrogen atoms by alkyl groups (rows in **Table 2.2a**) or when an alkyl group is substituted by a higher homologue (columns in **Table 2.2a**). When pure water as a medium is replaced by other solvent mixtures, known to solvate cations [6], mobilities change, but, with the exception of the small ammonium and methyl ammonium ions, the order of mobility magnitude within rows and columns remains the same. Some solvation cannot be excluded, as relative mobilities are not constant, but the effect is moderate enough so that ranking according to mobility is not disturbed.

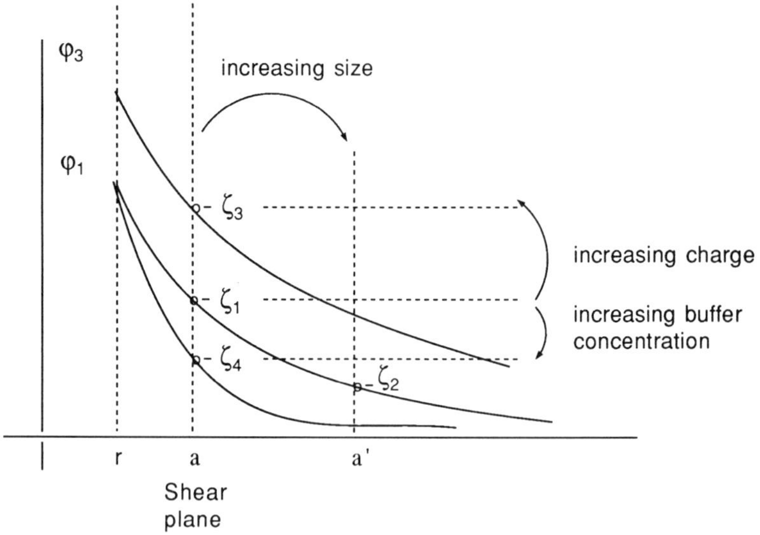

Figure 2.2
Effect of various parameters on the ζ potential.

Table 2.1 Mobility values [1] and ionic radius data [7] for some inorganic ions.

Cations	$\mu^{a)}$	r_{ion} $^{b)}$	r_{ion} $^{c)}$	Anions	$\mu^{a)}$
Li^+	38.7	0.86	3.40	F^-	-54.7
Na^+	50.5	1.12	2.76	Cl^-	-76.5
K^+	73.5	1.44	2.32	Br^-	-78.4
Rb^+	76.5	1.58	2.28	I^-	-77.0
Cs^+	78.0	1.84	2.28		

a) x 10^5 cm^2 / V.sec
b) Crystal radii, Å
c) Hydrated radii, Å

Table 2.2 Mobility values for alkyl ammonium ions (x 10^5 cm^2/V.sec) [6], pH = 5, (a) H_2O, (b) H_2O/MeOH (30/70), (c) H_2O/MeOH/DMSO (30/40/30).

	$R_n H_{4-n}N^+$	n = 0	n=1	n=2	n=3	n=4
(a)	R = H		72.2	72.2	72.2	72.2
	R = Me	72.2	56.5	53.1	47.6	44.9
	R = Et	72.2	48.6	37.9	33.8	32.9
	R = Pr	72.2	42.3	31.6	25.9	23.0
	R = Bu	72.2	39.0	27.2	22.0	18.5
(b)	R = H		**49.9**	**49.9**	49.9	49.9
	R = Me	**49.9**	51.3	51.1	49.0	46.5
	R = Et	49.9	45.6	40.8	36.5	37.8
	R = Pr	49.9	41.0	34.0	29.0	26.6
	R = Bu	49.9	37.7	29.4	23.3	22.4
(c)	R = H		**25.9**	**25.9**	**25.9**	**25.9**
	R = Me	**25.9**	**27.3**	29.2	27.5	26.4
	R = Et	25.9	25.5	24.5	22.4	23.9
	R = Pr	25.9	23.5	20.9	18.7	18.0
	R = Bu	25.9	22.3	18.7	16.0	14.9

Besides the ion's own properties, the medium also influences mobility. From eq.(2.3) and (2.4) it is obvious that mobility will depend on viscosity, as determined by temperature, cosolvents or other additives. To ascertain whether mobility changes are due solely to viscosity changes, or to neutralize the effect, it is common to consider the product of mobility and viscosity : the Walden product $\mu.\eta$ [6,8]. Due to the influence of the temperature on the viscosity, mobilities show a temperature dependence of about 2-3 %/° [9,10].

The effect of electrolytes, in zone electrophoresis supplied by the background buffer, can only be understood from the double-layer theory. The ions in the counter ion cloud cause the potential to decay exponentially. When the ionic strength I is increased, the potential will drop faster as is also obvious from the expression for the Debye length (double layer compression) [11] :

$$\delta^{-1} = \sqrt{\frac{\varepsilon \, . \, k.T}{4 \, . \, \pi \, . \, e^2} \frac{1}{\Sigma \, n_i z_i^2}} = \sqrt{\frac{\varepsilon \, . \, R.T}{8\, 10^3 \, . \, \pi \, . \, F^2 \, . \, I}} = \frac{0.3}{\sqrt{I}} \text{ (nm)} \tag{2.5}$$

with

$$I = \frac{1}{2} \sum_i c_i \, z_i^2 \tag{2.6}$$

When the buffer concentration, or more precisely, the ionic strength, is increased, a given potential φ_1 (**Figure 2.2**) will decay faster, resulting in a decreased potential ζ_4. By consequence, ion mobilities will decrease when background electrolyte concentration is increased. A theoretical relationship, based on the Debye length, has been established [11] :

$$\mu(I) = \mu(I{=}0) - \frac{z_i \, e \, \delta}{6 \, \pi \, \eta} = \mu(I{=}0) - f(\sqrt{I}) \tag{2.7}$$

but has not yet been experimentally verified within the context of capillary zone electrophoresis. It must be remembered that much of the theoretical framework on the double layer assumes relatively low electrolyte concentrations, less than 10 mM, whereas in capillary zone electrophoresis concentrations as high as 50 or 100 mM are often used.

2.1.2 Weak Electrolytes

For weak electrolytes the equilibrium between species with different charge states must be considered. Consider a monovalent base :

$$BH^+ \rightleftarrows B + H^+ \tag{2.8}$$

When the pH of the buffer is close to the pK_a value (± 2 units), the compound will occur in a partially charged state. When the exchange is fast, as is generally the case in acid-base equilibria, the fraction of the time spent in the charged state is equal to the dissociation constant α. Accordingly, the effective mobility μ_{eff} will only be a fraction of the absolute mobility μ_{abs}, i.e. the mobility that would be observed for the fully charged ion [1] :

$$\mu_{eff} = \alpha.\mu_{abs} \tag{2.9}$$

with, for proton acceptors (B / BH^+)

$$\alpha = \frac{10^{-pH}}{10^{-pH} + 10^{-pKa}} = \frac{1}{1 + 10^{pH - pKa}} \tag{2.10}$$

and, for proton donors (BH / B^-)

$$\alpha = \frac{10^{-pKa}}{10^{-pH} + 10^{-pKa}} = \frac{1}{1 + 10^{pKa - pH}} \tag{2.11}$$

By appropriately selecting the pH of the buffer, mobility differences can be induced between ions with similar absolute mobility (**Figure 2.3**) [12,13].

For multiple equilibria, absolute mobilities must be considered for each state [1] :

$$\mu_{eff} = \sum_i \alpha_i.\mu_{abs,i} \tag{2.12}$$

Complexation equilibria can be treated in an similar way, provided that the association and dissociation steps are fast.

Instead of considering different mobilities for each charged state, *Grossman* et al. have developed a peptide mobility model [14], based on eq.(2.3) and a calculated effective charge :

$$q_{eff} = f\left(\alpha_i, q_i\right) \tag{2.13}$$

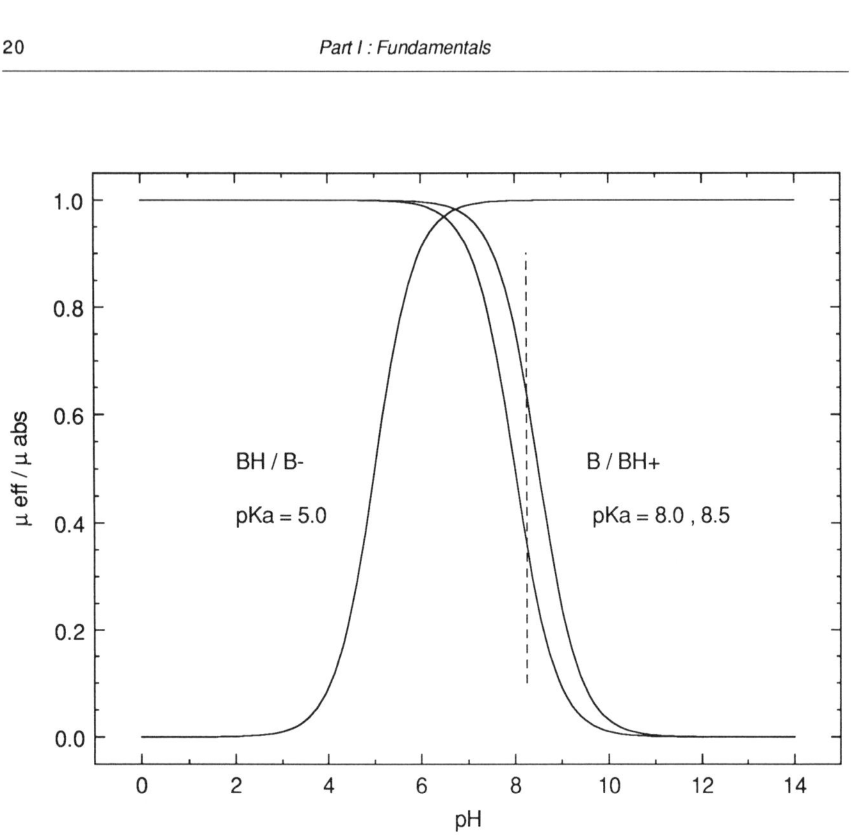

Figure 2.3
Ratio of effective to absolute mobility for weak electrolytes as a function of buffer pH.

2.2 Electroosmosis

2.2.1 Electroosmotic Flow

The charge separation between wall and electrophoretic medium, giving rise to electroosmosis, also results in an electric double layer with its own ζ potential (**Figure 2.4**). The relation between ζ potential and electroosmotic mobility is very similar to the one for ions, differing only by a proportionality factor which reflects the transition from curved to flat geometry :

$$\mu_{eo} = \frac{\varepsilon \,.\, \zeta}{4\,\pi\,\eta} = \frac{\varepsilon_0 \,.\varepsilon_r \,.\, \zeta}{\eta} \tag{2.14}$$

As with electrophoretic velocity, the observed electroosmotic velocity is proportional to the applied electric field :

$$v_{eo} = \mu_{eo} \cdot E \tag{2.15}$$

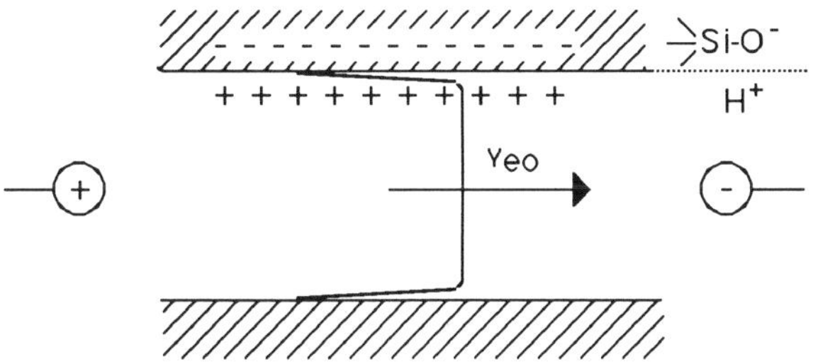

Figure 2.4
Charge separation at column-buffer interface and electroosmotic flow.

The direction is determined by the charge of the excess ions in the solution, originating from the charge separation process. The acidic silanol groups on glass or fused silica surfaces will loose protons (positive charge excess in the solution) causing an electroosmotic flow into the direction of the cathode. When long-chain ammonium salts are added to the buffer, they can be adsorbed by the surface [15], leaving an excess of counter-anions in the buffer and, as a consequence, reversing the direction.

Several methods have been used to measure electroosmotic flow in capillaries, some of which are only practical with relatively wide-bore capillaries :
- addition of an UV-active marker to one electrode vessel and monitoring the absorbance in the other [16,17],
- direct observation of volume or level increase in the destination vessel [16],
- weighing the mass of buffer transferred [17-19],
- monitoring electric current while a buffer with different conductivity is drawn into the capillary by the electroosmotic flow [20],
- determination of ζ_{wall} from streaming potential measurements [19],
- monitoring the migration time of a neutral marker [21,22].

The last method is very convenient, and for most purposes satisfactory. It has the advantages that it can be used to control electroosmotic flow during analysis.

With a UV detector, dimethylformamide, mesityloxide or another UV-absorbing substance can be used. Even non-absorbing species, like methanol, cause baseline disturbances that can be used as a marker.

2.2.2 The Electroosmotic Flow Profile

Strictly, the electroosmotic flow is a function of the distance from the wall and, with $\varphi(x)$ representing the potential between wall and buffer (similar as in **Figure 2.1**), eq.(2.15) must be rewritten as [10]

$$v_{eo}(x) = \frac{\varepsilon}{4\pi\eta}(\zeta - \varphi(x))\,E = \frac{\varepsilon}{4\pi\eta}(\zeta - \varphi(x{=}0).e^{-\delta x})\,E \tag{2.16}$$

Using a more rigorous treatment, *Rice* and *Whitehead* have shown [23] that the flow does not vary over 90% of the capillary diameter when $\kappa.r_c > 50$, or, using eq.(2.5)

$$r_c\,(\mu m)\,\sqrt{I\,(mol/l)} > 0.015 \tag{2.17}$$

Even for a very low ionic strength (10^{-3} M) and for a small capillary diameter (2 μm) this condition is fulfilled, justifying the essentially flat or plug-like flow as represented in **Figure 2.4**.

If the ratio of capillary radius versus Debye length is unfavorable (a condition known as 'double layer overlap'), the velocity will vary along the radial position, in the worst case approaching a parabolic-like profile, as with pressure-driven flow. The nature of the profile has consequences for the zone spreading, as will be discussed later.

2.2.3 Influence of Buffer pH and Concentration on the Electroosmotic Flow

According to the double layer theory, the response of the electroosmotic mobility to changes in ionic strength or charge, or in this case more precisely the surface charge density, is similar to that described above for the electrophoretic mobility. In the case of glass or fused silica capillaries the surface charge density, originating from the dissociation of surface silanol groups, is influenced by the pH of the buffer.

Electroosmotic mobilities, or equivalent linear velocities and migration times, have been related to buffer concentration [8,18,24-28] and ionic strength [26,28]. *VanOrman* et al. have demonstrated that ionic strength is a more meaningful quantity than concentration [28] (**Figure 2.5**). The studies confirm that, at a fixed pH, mobility decreases with increasing concentration, but confusion exists about the mathematical model that should be used to describe this behaviour.

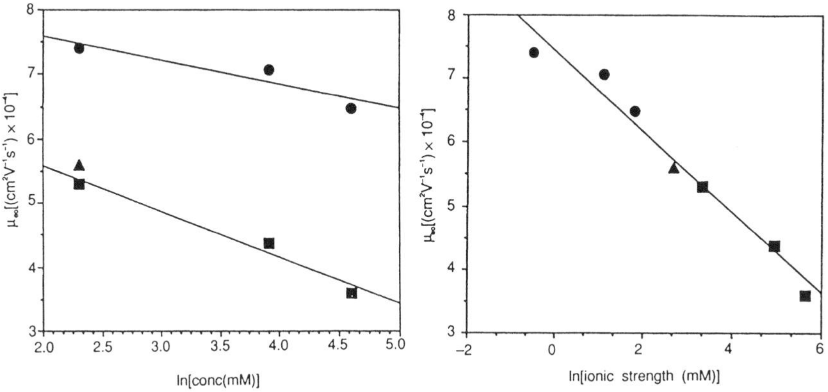

Figure 2.5
Electroosmotic mobility in a fused silica capillary versus natural logarithm of buffer concentration (left) and ionic strength (right) at pH 8.0. Circle = borate buffer, triangle = carbonate buffer, square = phosphate buffer. *(Reprinted with permission [28], © MicroSeparations Inc.)*

The pH-dependence of the electroosmotic mobility in fused silica capillaries has been determined [18,21,29-33] (**Figure 2.6**). Observed values are variable, influenced by the buffer system used, but also by brand [34] and history [33] of the fused silica tubing. The electroosmotic mobility is very low in strongly acidic buffers, starts to rise above a pH of 3, and increases strongly between a pH of 4 and pH values of 7 to 8. At higher pH, mobility still increases, but more modestly [29,35].

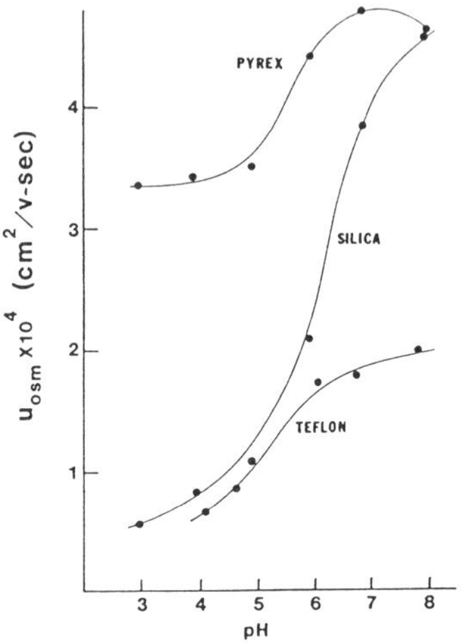

Figure 2.6
Effect of pH on electroosmotic mobility in different capillaries (at constant ionic strength). *(Reprinted with permission [29], © Hüthig Verlag)*

2.3 Observed or Apparent Ion Mobility

The apparent [22] or observed mobility results from the vectorial combination of electrophoretic and electroosmotic mobility :

$$\mu_{app} = \mu_{ep} + \mu_{eo} \tag{2.18}$$

and

$$v_{app} = \mu_{app}.E = (\mu_{ep} + \mu_{eo}).E = v_{ep} + v_{eo} \tag{2.19}$$

All possible combinations have been summarized in **Table 2.3**. In the case of a fused silica capillary, using a neutral or alkaline buffer, cations will move to the cathode with a velocity greater than their pure electrophoretic mobility would allow. Neutral components will migrate in a single zone with a velocity determined by the electroosmotic flow. Finally, also anions will move to the cathode, their velocity decreasing with increasing mobility, until their mobility is large enough to counteract the electroosmotic flow in the opposite direction.

Everaerts has labeled these migration modes as downstream, midstream and upstream respectively [36]. Regardless of the direction of the electroosmotic flow, ions in downstream mode have always a charge opposite to that of the column wall, and increased risk of being adsorbed. Ions in the upstream mode, however, are charged similarly to the wall and are repelled. As will be discussed later, this type of wall interaction has consequences for the quality and the efficiency of analytical applications.

The many factors that play a role in the determination of the observed velocity are summarized in **Figure 2.7**.

Table 2.3 Influence of electroosmotic flow on the direction of ion movement.

		v_{eo}	v_{ep}	v_{tot}	cations move to ...	
$\mu_{eo} > 0$		+	+	+	cathode	(accelerated)
$\mu_{eo} = 0$		0	+	+	cathode	(pure electrophoresis)
$\mu_{eo} < 0$	$\lvert\mu_{eo}\rvert < \lvert\mu_{ep}\rvert$	-	+	+	cathode	(decelerated)
$\mu_{eo} < 0$	$\lvert\mu_{eo}\rvert > \lvert\mu_{ep}\rvert$	-	+	-	anode	(reversed movement)
		v_{eo}	v_{ep}	v_{tot}	anions move to ...	
$\mu_{eo} > 0$	$\lvert\mu_{eo}\rvert > \lvert\mu_{ep}\rvert$	+	-	+	cathode	(reversed movement)
$\mu_{eo} > 0$	$\lvert\mu_{eo}\rvert < \lvert\mu_{ep}\rvert$	+	-	-	anode	(decelerated)
$\mu_{eo} = 0$		0	-	-	anode	(pure electrophoresis)
$\mu_{eo} < 0$		-	-	-	anode	(accelerated)

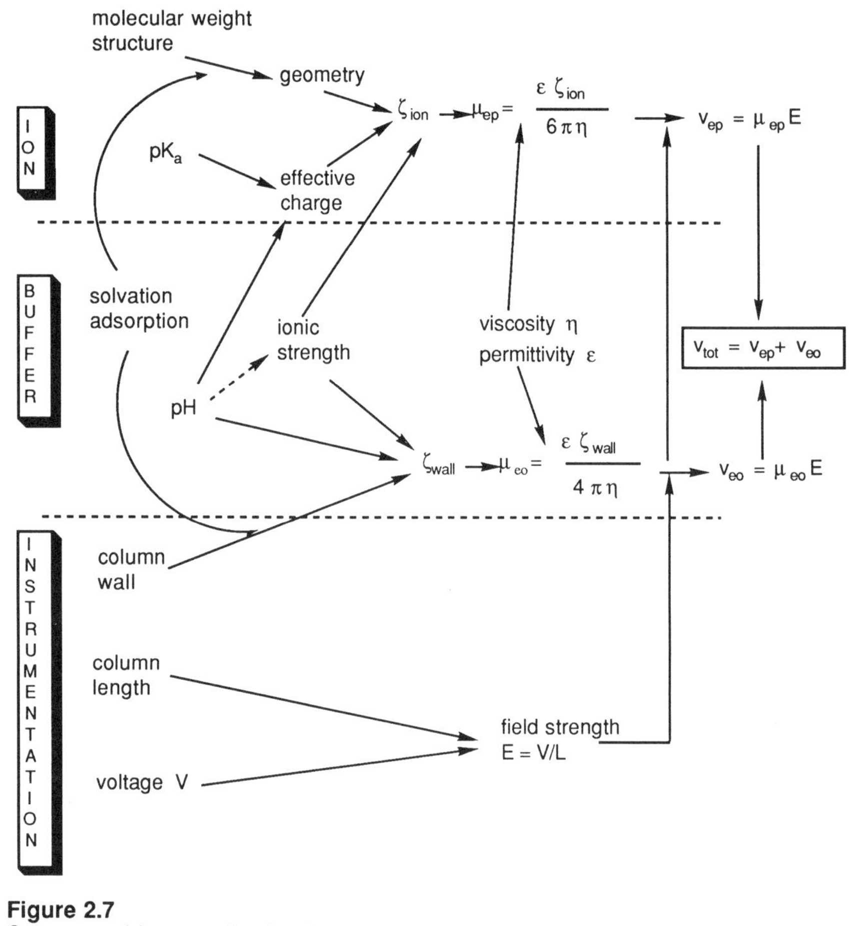

Figure 2.7
Summary of factors affecting the observed velocity of ions.

References

[1] F.M.Everaerts, J.L.Beckers, Th.P.E.M.Verheggen, "Isotachophoresis : Theory, Instrumentation and Applications", Elsevier Science Publ., Amsterdam, 1976 (J.Chromatogr.Lib., vol.6).

[2] D.Stigter, K.J.Mysels, J.Phys.Chem., **59** (1955) 45.
Tracer electrophoresis. II. The mobility of the micelle of sodium lauryl sulfate and its interpretation in terms of zeta potential and charge.

[3] D.C.Grahame, Chem.Rev., **41** (1947) 441.
The electrical double layer and the theory of electrocapillarity.

[4] S.Hjertén, Topics Bioelectrochem. Bioenerg., **2** (1978), 89.
Analytical electrophoresis.

[5] B.M.Michov, Electrophoresis (Weinheim), **9** (1988) 201.
A quantitative theory of the Stern electric double layer.

[6] E.Kenndler, P.Jenner, J.Chromatogr., **390** (1987) 185.
Isotachophoresis in mixed solvents consisting of water, methanol and dimethyl sulphoxide. II. Solvation effects on the acid-base and transport properties of cation acids of the ammonium type.

[7] F.A.Cotton, G.Wilkinson, "Advanced Inorganic Chemistry, A comprehensive Text", 3rd ed., Interscience Publ., New York, 1972.

[8] D.S.Burgi, K.Salomon, R.-L.Chien, Poster presented at the 2nd International Symposium on High Performance Capillary Electrophoresis, San Francisco, January 29-31,1990 (P-112).
The electro-osmotic pump : ways to tame the beast.

[9] J.W.Jorgenson, K.D.Lukacs, Anal.Chem., **53** (1981) 1298.
Zone electrophoresis in open-tubular glass capillaries.

[10] S.Hjertén, Chromatogr.Rev., **9** (1967) 122.
Free zone electrophoresis.

[11] B.M.Michov, Electrophoresis (Weinheim), **6** (1985) 471.
Ionic mobility parameter.

[12] S.Terabe, T.Yashima, N.Tanaka, M.Araki, Anal.Chem., **60** (1988) 1673.
Separation of oxygen isotopic benzoic acids by capillary zone electrophoresis based on isotope effects on the dissociation of the carboxyl group.

[13] S.Wren, J.Microcol.Sep., **3** (1991) 147.
Optimization of pH in the electrophoretic separation of 2-, 3-, and 4-methylpyridines.

[14] P.D.Grossman, J.C.Colburn, H.H.Lauer, Anal.Biochem., **179** (1989) 28.
A semiempirical model for the electrophoretic mobilities of peptides in free solution capillary electrophoresis.

[15] T.Tsuda, J.High Resolut.Chromatogr. Chromatogr.Commun., **10** (1987) 622.
Modification of electroosmotic flow with cetyltrimethylammonium bromide in capillary zone electrophoresis.

[16] T.S.Stevens, H.J.Cortes, Anal.Chem., **55** (1983) 1365.
Electroosmotic propulsion of eluent through silica based chromatographic media.

[17] B.J.Wanders, A.A.A.M.Van de Goor, F.M.Everaerts, J.Chromatogr., **470** (1989) 89.
Methods for on-line determination and control of electroendosmosis in capillary electrochromatography and electrophoresis.

[18] K.D.Altria, C.F.Simpson, Chromatographia, **24** (1987) 527.
High voltage capillary zone electrophoresis : Operating parameters effects on electroendosmotic flows and electrophoretic mobilities.

[19] A.A.A.M.Van de Goor, B.J.Wanders, F.M.Everaerts, J.Chromatogr., **470** (1989) 95.
Modified methods for off- and on-line determination of electroosmosis in capillary electrophoretic separations.

[20] X.Huang, M.J.Gordon, R.N.Zare, Anal.Chem., **60** (1988) 1837.
Current-monitoring method for measuring the electroosmotic flow rate in capillary zone electrophoresis.

[21] T.Tsuda, K.Nomura, G.Nakagawa, J.Chromatogr., **264** (1983) 385.
Separation of organic and metal ions by high-voltage capillary electrophoresis.

[22] J.L.Beckers, F.M.Everaerts, M.T.Ackermans, J.Chromatogr., **537** (1991) 407.
Determination of absolute mobilities, pK values and separation numbers by capillary zone electrophoresis. Effective mobility as a parameter for screening.

[23] C.L.Rice, R.Whitehead, J.Phys.Chem., **69** (1965) 4017.
Electrokinetic flow in a narrow cylindrical capillary.

[24] S.Fujiwara, S.Honda, Anal.Chem., **58** (1986) 1811.
Determination of cinnamic acid and its analogues by electrophoresis in a fused silica capillary tube.
[25] G.J.M.Bruin, J.P.Chang, R.H.Kuhlman, K.Zegers, J.C.Kraak, H.Poppe, J.Chromatogr., **471** (1989) 429.
Capillary zone electrophoretic separations of proteins in polyethylene glycol-modified capillaries.
[26] V.Dolnik, J.Liu, F.Banks, M.V.Novotny, P.Boček, J.Chromatogr., **480** (1989) 321.
Capillary zone electrophoresis of oligonucleotides. Factors affecting separations.
[27] K.Salomon, D.S.Burgi, J.S.Helmer, Poster presented at the 2nd International Symposium on High Performance Capillary Electrophoresis, San Francisco, January 29-31,1990 (P-306).
Control of ion mobilities and electroosmotic flow in the separation of a pharmaceutical sample.
[28] B.B.VanOrman, G.G.Liversidge, G.L.McIntire, T.M.Olefirowicz, A.G.Ewing, J.Microcol. Sep., **2** (1990) 176.
Effects of buffer composition on electroosmotic flow in capillary electrophoresis.
[29] K.D.Lukacs, J.W.Jorgenson, J.High Resolut.Chromatogr.Chromatogr.Commun., **8** (1985) 407.
Capillary zone electrophoresis : Effect of physical parameters on separation efficiency and quantitation.
[30] R.M.McCormick, Anal.Chem., **60** (1988) 2322.
Capillary zone electrophoretic separation of peptides and proteins using low pH buffers in modified silica capillaries.
[31] R.D.Smith, H.R.Udseth, J.A.Loo, B.W.Wright, G.A.Ross, Talanta, **36** (1989) 161.
Sample introduction and separation in capillary electrophoresis, and combination with mass spectrometric detection.
[32] K.Otsuka, S.Terabe, J.Microcol.Sep., **1** (1989) 150.
Effect of pH on electrokinetic velocities in micellar electrokinetic chromatography.
[33] W.J.Lambert, D.L.Middleton, Anal.Chem., **62** (1990) 1585.
pH hysteresis effect with silica capillaries in capillary zone electrophoresis.
[34] S.Terabe, Y.Miyashita, O.Shibata, E.R.Barnhart, L.R.Alexander, D.G.Patterson, B.L.Karger, K.Hosoya, N.Tanaka, J.Chromatogr., **516** (1990) 23.
Separation of highly hydrophobic compounds by cyclodextrin-modified micellar electrokinetic chromatography.
[35] J.Vindevogel, P.Sandra, J.Chromatogr., **541** (1991) 483.
Simultaneous pH and ionic strength effects and buffer selection in capillary electrophoretic techniques.
[36] F.M.Everaerts, A.A.A.M.Van de Goor, Th.P.E.M.Verheggen, J.L.Beckers, J.High Resolut. Chromatogr.Chromatogr.Commun., **12** (1989) 28.
Electrophoresis versus electrochromatography.

3 Capillary Zone Electrophoresis (CZE)

3.1 Electropherograms : Migration Times, Efficiency and Resolution

3.1.1 Migration Times

Migration times can be calculated from the relationships between mobility, velocity, applied voltage and column length [1] :

$$t_M = \frac{L_d}{v} = \frac{L_d}{\mu E} = \frac{L_d L_V}{\mu V} \tag{3.1}$$

When the difference between L_d (column length between inlet and detector) and L_V (column length between inlet and outlet) is ignored or negligible, eq.(3.1) reduces to

$$t_M = \frac{L^2}{\mu V} \tag{3.2}$$

For a given applied voltage, migration times will increase with the square of the column length (**Figure 3.1**), hence the importance of using short columns in combination with high voltages.

As the mobility parameter in eq.(3.1) includes the effect of electroosmosis, determination of electrophoretic mobility from electropherograms requires the measurement of t_0, the electroosmotic breakthrough time :

$$t_0 = \frac{L_d L_V}{\mu_{eo} V} \tag{3.3}$$

and, considering that $\mu = \mu_{eo} + \mu_{ep}$:

$$\mu_{ep} = \frac{L_d L_V}{V}\left(\frac{1}{t_M} - \frac{1}{t_0}\right) = \frac{L_d L_V (t_0 - t_M)}{V t_0 t_M} \tag{3.4}$$

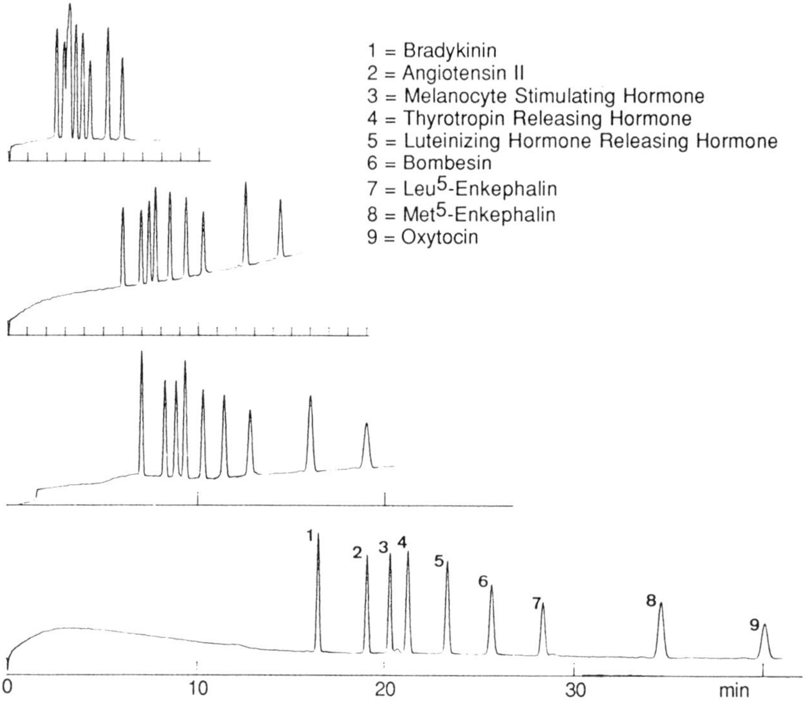

Figure 3.1
Effect of column length on analysis time : analysis of peptides. Detection : UV 200 nm. Columns[1] : polyacrylamide coated fused silica, (top to bottom) 25 μm x 18/20 cm, 25 μm x 28/30 cm, 50 μm x 33/35 cm, 50 μm x 48/50 cm. Applied voltage 10 kV. Buffer : 100 mM phosphate, pH 2.5.

3.1.2 Efficiency

Under ideal conditions, the only factor contributing to peak dispersion is axial diffusion of the ions which is expressed by Einstein's equation :

$$\sigma_L^2 = 2\, D_m\, t \tag{3.5}$$

---

[1] *In figure captions, colum dimensions are specified as d_c x L_d/L_V (or d_c x L if L_d and L_V are not separately specified). Negative voltages (or currents) indicate reversed polarity, i.e. cathode at the injection side.*

with, the diffusion coefficient D_m determined by

$$D_m = \frac{k\,T}{6\,\pi\,\eta\,r_{eff}} \left(\frac{m^2}{sec}\right) \tag{3.6}$$

After identifying the column residence time t with t_M, and using the familiar equations for plate height h and plate count N, eq.(3.5) leads to :

$$h = \frac{\sigma_L^2}{L_d} = \frac{2\,D_m}{L_d}\,\frac{L_d\,L_V}{\mu\,V} = \frac{2\,D_m\,L_V}{\mu\,V} \tag{3.7}$$

and

$$N = \frac{L_d}{h} = \frac{\mu\,V\,L_d}{2\,D_m\,L_V} \tag{3.8}$$

or, in its simplified form ($L_d \approx L_V$) [2,3] :

$$N = \frac{\mu\,V}{2\,D_m} \tag{3.9}$$

From eq.(3.9) it follows that high efficiency will be obtained by applying high voltages, i.e. by limiting the residence time in the column. Efficiency will also be highest for large ions like proteins, as they are less diffusive ($D_m \approx 10^{-10}$ m^2/sec) than small molecules ($D_m \approx 10^{-9}$ m^2/sec). It is interesting to note that those analytes that would show reduced efficiency in HPLC (through the influence of D_m on mass transfer) perform best in CZE.

Depending on the actual values of D_m, μ and V, calculated maximum values of N are in the range 10^5 - 10^6. However, due to non-diffusional band broadening (deviation from ideal behaviour), lower efficiencies, although still impressive when compared to HPLC, will generally be observed.

3.1.3 Resolution

By analogy with chromatography, resolution R_s between closely eluting peaks is defined as the difference in migration time, relative to the average peak width at base :

$$R_S = \frac{\Delta t_M}{\frac{1}{2}(w_1 + w_2)} = \frac{\Delta t_M}{\overline{w}} \tag{3.10}$$

Assuming that $\overline{\mu} \approx \mu_1 \approx \mu_2$, the migration time difference can be written as :

$$\Delta t_M = \frac{L_d . L_V}{V}\left(\frac{\Delta\mu}{\overline{\mu}^2}\right) \tag{3.11}$$

From

$$N = \frac{t_M^2}{\sigma^2} = \frac{t_M^2}{(\frac{w}{4})^2} \tag{3.12}$$

and using eq.(3.1), it follows that the average peak width at base is given by

$$\overline{w} = \frac{4.\overline{t_M}}{\sqrt{N}} = 4\,\frac{L_d\,L_V}{\overline{\mu}\,V}\sqrt{\frac{1}{N}} \tag{3.13}$$

or [4]

$$R_S = \frac{\Delta\mu}{\overline{\mu}} . \frac{\sqrt{N}}{4} \tag{3.14}$$

Under ideal conditions, N can be substituted, using eq.(3.8), which results in :

$$R_S = \frac{1}{4\sqrt{2}}\,\Delta\mu\sqrt{\frac{V}{D_m\,\overline{\mu}} . \frac{L_d}{L_V}} \tag{3.15}$$

or in its simplified form ($L_d \approx L_V$) [1] :

$$R_S = \frac{1}{4\sqrt{2}}\,\Delta\mu\sqrt{\frac{V}{D_m\,\overline{\mu}}} \qquad \left(\frac{1}{4\sqrt{2}} = 0.177\right) \tag{3.16}$$

Let it be remembered that

$$\Delta\mu = \Delta\mu_{ep} \tag{3.17}$$

and

$$\overline{\mu} = \overline{\mu_{ep}} + \mu_{eo} \tag{3.18}$$

The influence of total mobility, mobility differences and applied voltage is shown in **Figure 3.2** and compared to the resolution function in HPLC. Increasing the applied voltage improves resolution, comparable to efficiency in HPLC, but the square root relation tends to level off the effect. As will be shown, the possible increase of the applied voltage is further limited by Joule heating, invalidating eq.(3.8) and by consequence also eq.(3.16). By analogy to the selectivity factor α in HPLC, inducing mobility differences is an effective way to improve resolution.

When the average electrophoretic mobility becomes equal, but opposite in sign, to the electroosmotic mobility (eq.(3.18)), resolution tends to infinity. But the same will be true for the analysis time. Exploitation of the increase in resolution requires that the system already shows some resolving power (**Figure 3.3**). Using eq.(3.1) and eq.(3.16), it can be shown that increasing R_s from 0.5 (onset of visible peak splitting) to 1.5 (baseline separation), solely by decreasing the average mobility, will increase the analysis time by a factor 9.

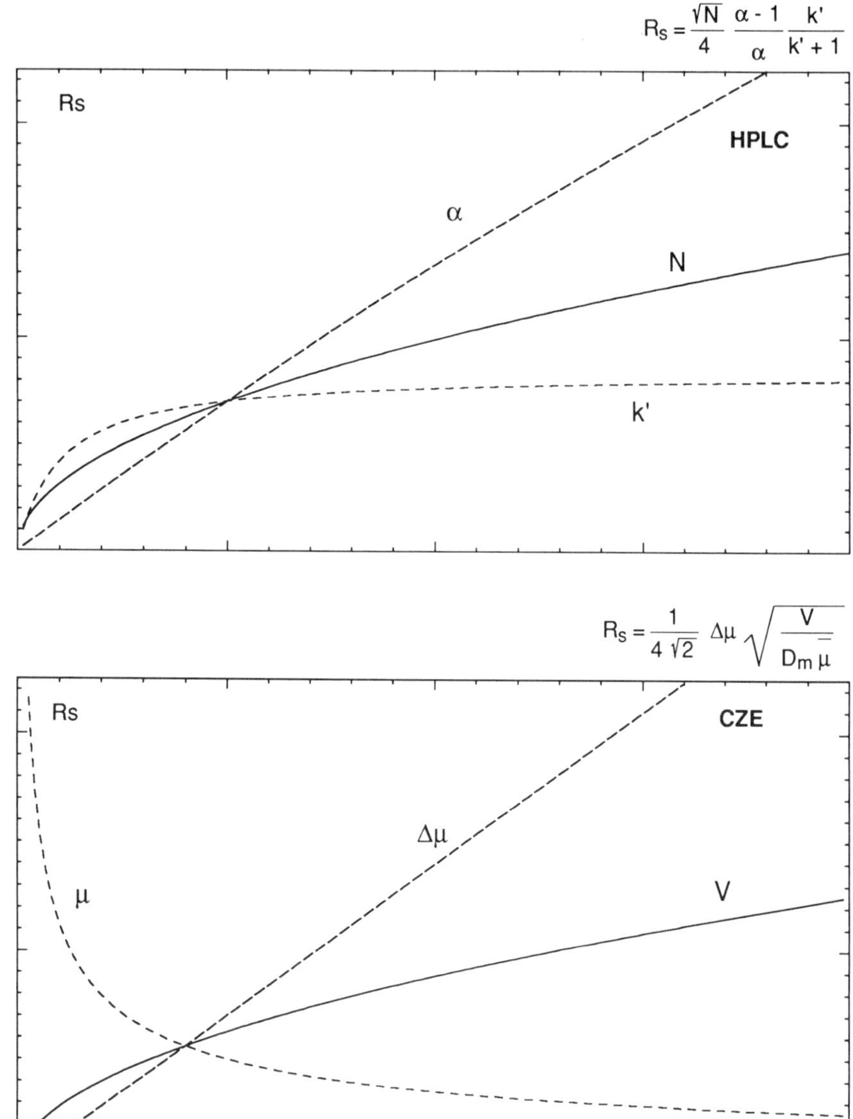

Figure 3.2
Factors affecting resolution in HPLC and CZE.

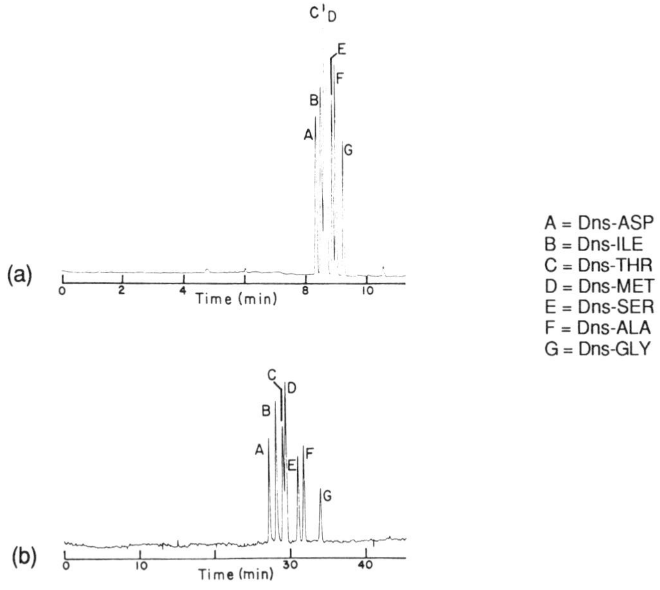

Figure 3.3
Effect of average mobility on resolution. (a) Untreated capillary. (b) TMCS-treated capillary with reduced electroosmotic flow. *(Reprinted with permission [2] , © American Chemical Society)*

3.1.4 The Plate Number Concept in Zone Electrophoresis

> " ... *nothing resembling real plates or stages exists (...) the plate model confuses the true mechanism of such systems."* [5]
>
> " *We do not understand the significance of all these numbers ...* " [6]

The fundamental process underlying separation efficiency is peak dispersion. The use of the theoretical plate concept to describe peak spreading in chromatography has been criticized [4,5], but was found justified for practical reasons, providing a parameter that characterizes column performance. Accordingly, it has been introduced in electrophoresis in order to compare its separation power to that of chromatography [7] although there is nothing in the column that suggests discrete stages.

When CZE is compared to chromatography, ignoring tacit assumptions can lead to confusing conclusions. According to eq.(3.9), N is "somewhat surprisingly" [2] independent of column length, whereas in chromatography increasing column length is a common practice to increase the available plate number. This is not a mysterious property of zone electrophoresis but reflects operational differences : '*changing the column length, while leaving everything else as it was*' has a different meaning. In the case of chromatography, the instrumental parameter is the flow rate which is also the driving force. In CZE however, the instrumental parameter V must be changed to keep the driving force E constant. Rewriting eq.(3.8) or (3.9) accordingly results in :

$$N = \frac{\mu\, L_d\, E}{2\, D_m} \tag{3.19}$$

reintroducing column length. A similar observation can be made in relation to the resolution equation (3.15).

In uniform chromatography all peaks move with the same speed in the mobile phase when eluting from the column. The resulting constant ratio between length and time-based properties makes comparison between peaks straightforward. For instance, eq.(3.10) can be interpreted in length units, as well as in time units [4]. But in zone electrophoresis each peak moves with its own characteristic velocity through the detector [8]. Properties measured on a chart paper or a digitized electropherogram are biased by this fact. Depending upon its end-use, the observed peak spreading σ_t may need correction [9] according to :

$$\sigma_L = \sigma_t . v = \sigma_t \frac{L_d}{t_M} \tag{3.20}$$

3.2 Non-Ideal Behaviour and Loss of Efficiency

Theoretically, capillary zone electrophoresis shows all the characteristics necessary for fast separations. Applying a high voltage over a short column, reduces analysis time (eq.(3.1)) and improves efficiency (eq.(3.8)) simultaneously. This is contrary to chromatography, where, according to the van Deemter equation, the flow rate cannot be reduced below a critical value without sacrificing separation efficiency.

In spite of the remarkable achievements, like 3-400,000 theoretical plates generated within 70 seconds (separation of 8 labeled amino acids in a 35 cm x 10 µm ID column) [10] (**Figure 3.4**), observed efficiency is generally lower than predicted. This is indicative of contributions to zone spreading other than axial diffusion alone. Several factors have been identified, the details of which will be discussed in subsequent chapters on thermal effects and instrumental aspects.

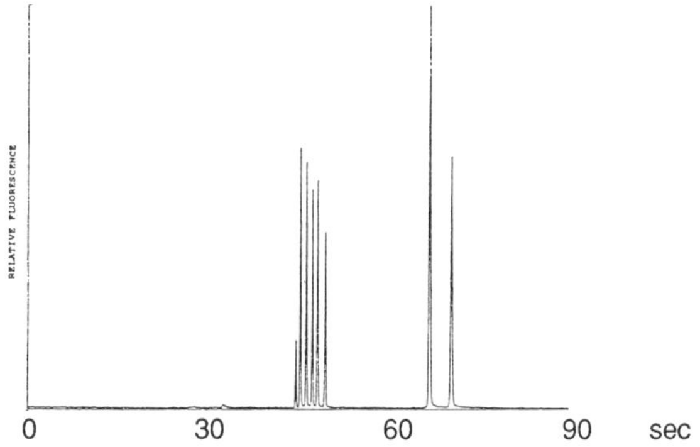

Figure 3.4
Fast CZE separation and laser-induced fluorescence detection of 8 NDA-amino acids. (10 µm x 35 cm, 30 kV, 10 mM borate/20 mM KCl, pH 9.5). The elution order is Trp, Phe, Val, Ser, Ala, Gly, Glu, Asp). *(Reprinted with permission [10], © Hüthig Verlag)*

Applied voltage, column length, column diameter and background buffer conductivity all influence energy dissipation (Joule heating). Although this was the very reason for miniaturization of electrophoresis, it was soon realized [2] that its effects were not completely banned and impose upper limits on the voltage gradients that can be used.

Inside the column, interaction of the analyte with the wall (reversible adsorption) will induce mass transfer-related peak broadening. At high concentrations the local voltage gradient can be disturbed, leading to migrational dispersion (sample overloading). Finally, extra-column zone broadening, occurring in injection and detection systems, must be considered.

A more comprehensive treatment of efficiency [8,11,12], including these effects,

$$\sigma_{tot}^2 = \sigma_D^2 + \sigma_{\Delta T}^2 + \sigma_{inj}^2 + \sigma_{det}^2 + \sigma_{ads}^2 + \sigma_{migr}^2 + \dots \qquad (3.21)$$

reintroduces the concept of an optimal, van Deemter-like velocity, or voltage to column length ratio.

3.3 Possible Pitfalls in Resolution Optimization

Apart from the non-ideal behaviour of the efficiency, the application of eq.(3-16) for resolution optimization is further complicated by cooperative effects, already indicated in **Figure 2.7**. For instance, when pH-adjustment or complexation is used to optimize mobility differences ($\Delta\mu$), the changes in the buffer composition can also affect the electroosmotic mobility, included in the average mobility parameter in the resolution equation. This effect can be cooperative or counteractive in obtaining the desired resolution increase.

Also, changing pH can influence electroosmotic mobility in more than one way. Under certain conditions, an unexpected decrease in the electroosmotic flow can be observed when the the pH is increased at constant buffer concentration [13,14]. Several examples can be found in the literature [14]. Ascribing this phenomenon to a 'decrease in the ζ-potential with pH' [15] is just a restatement of the same fact as no significant changes in dielectric constant or viscosity (eq.(2.14)) have to be expected. This behaviour is assumed to be related to changes in the ionic strength caused by the titrant during pH adjustment. This hypothesis has been tested by measuring the electroosmotic mobilities in different buffers [14], using mesityl oxide as a marker. Four buffer systems were used :

- TRIS/HCl (TRIS = tris(hydroxymethyl)-aminomethane)
- H_3BO_3/NaOH
- $Na_2B_4O_7$/NaOH
- $Na_2B_4O_7$/HCl

The results (**Figure 3.5**) confirm that boric acid and borax are functionally similar when titrated with a base [16,17] :

$$Na_2B_4O_7 + 7\ H_2O \rightleftarrows 2\ Na^+\ B(OH)_4^- + 2\ H_3BO_3 \rightleftarrows 2\ NaOH + 4\ H_3BO_3$$

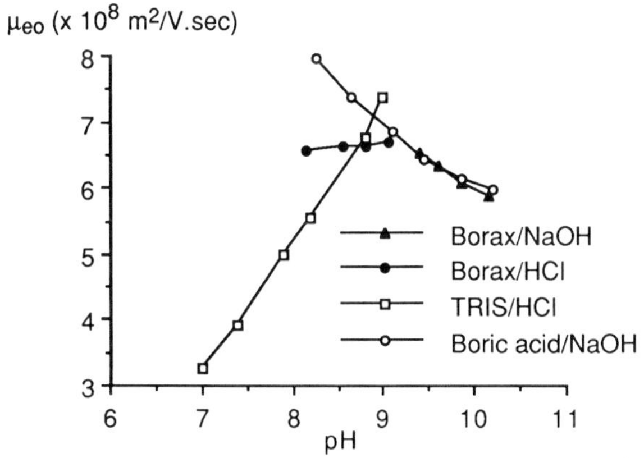

Figure 3.5
Influence of buffer pH on electroosmotic mobility in fused silica capillaries. Buffer concentration : 50 mM. *(Reprinted with permission [14], © Elsevier Science Publishers B.V.)*

For this reason, borax-concentrations have been expressed here as their boric acid equivalent (e.g. 25 mM borax is equivalent to 100 mM boric acid).

Three different cases remain to relate pH and ionic strength during titration, assuming that for these weak electrolytes, the ion concentrations are equal to the analytical concentrations of the titrant :

(a) TRIS + HCl $\rightleftarrows$ $TRISH^+ + Cl^-$

(b) H_3BO_3 + NaOH $\rightleftarrows$ $Na^+ + B(OH)_4^-$

(c) $2\ Na^+ + 2B(OH)_4^- + HCl \rightleftarrows 2\ Na^+ + B(OH)_4^- + Cl^- + H_3BO_3$

In case (a) a decrease in pH is accompanied by an increase of the ionic strength. Both effects are cooperative in producing a relatively strong dependence of the electroosmotic mobility on the pH. In case (b) however, increasing the pH will result in an increase of the ionic strength. Apparently, the ionic strength effect is stronger than the 'pure' pH-effect resulting in an inverted dependence of the electroosmotic flow on the pH. The borax/NaOH-case is similar as it can be considered as a boric acid/NaOH-system that has already been partially neutralized. All the cited examples that show lower electroosmotic mobility at higher pH belong to this category. Case (c) is interesting; the amount of ions remains constant during neutralization. Data confirm that mobility does indeed increase slightly with pH, as expected. In this pH-range however, the effect is so moderate that it is easily overcompensated by ionic strength effects.

When differences in the amount of titrant are compensated by adding a calculated amount of salt, the anomalous decrease in mobility with pH disappears and a rather flat profile is obtained (**Figure 3.6**). The confusion between constant concentration and constant ionic strength explains the anomaly. In existing studies, the use of a pH scale with fixed ionic strength is exceptional. In most of the reports, where the pH change is limited to a few units only, constant concentration buffers are used, often by mixing two constituents, e.g. primary and secondary phosphate, in varying ratios. There may be valid practical reasons to do so, but a numerical example demonstrates the possible pitfall of such a procedure. A 100 mM borate buffer, 5% neutralized with NaOH (considering the first neutralization step only) has a pH of 8.1 and an ionic strength of 5 mM. Doubling the concentration will double the ionic strength. However, increasing the pH of the 100 mM solution by 1 unit will require a degree of neutralization of ca. 24 %, resulting in an ionic strength of 24 mM, nearly a five-fold increase. This explains why ionic strength effects can be more prominent through pH than through buffer concentration.

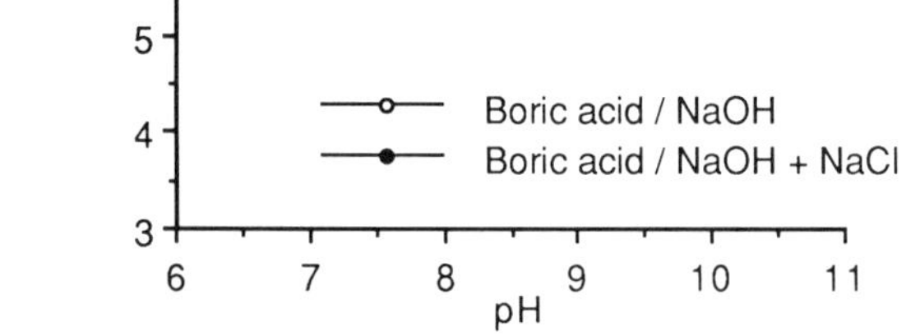

Figure 3.6
Influence of added salt on electroosmotic mobility : Constant concentration (50 mM) vs. constant ionic strength (50 mM). *(Reprinted with permission [14], © Elsevier Science Publishers B.V.)*

References

[1] J.W.Jorgenson, K.D.Lukacs, J.High Resolut.Chromatogr.Chromatogr.Commun., **4** (1981) 230.
Zone electrophoresis in open-tubular glass capillaries : Preliminary data on performance.

[2] J.W.Jorgenson, K.D.Lukacs, Anal.Chem., **53** (1981) 1298.
Zone electrophoresis in open-tubular glass capillaries.

[3] J.W.Jorgenson, K.D.Lukacs, J.Chromatogr., **218** (1981) 209.
High-resolution separations based on electrophoresis and electroosmosis.

[4] J.C.Giddings, "Dynamics of Chromatography. Part I : Principles and Theory", Dekker Inc, New York, 1965.

[5] J.C.Giddings, J.Chromatogr., **395** (1987) 19.
Transport, space, entropy, diffusion and flow. Elements underlying separation by electrophoresis, chromatography, field-flow fractionation and related methods.

[6] E.Wenisch, C.Tauer, A.Jungbauer, H.Katinger, M.Faupel, P.G.Righetti, J.Chromatogr., **516** (1990) 133.
Capillary zone electrophoresis for monitoring r-DNA protein purification in multicompartment electrolyzers with immobiline membranes.

[7] J.C.Giddings, Sep.Sci., **4** (1969) 181.
Generation of variance, "theoretical plates", resolution and peak capacity in electrophoresis and sedimentation.

[8] X.Huang, W.F.Coleman, R.N.Zare, J.Chromatogr., **480** (1989) 95.
Analysis of factors causing peak broadening in capillary zone electrophoresis.

[9] S.Hjertén, K.Elenbring, F.Kilár, J.-L.Liao, A.J.C.Chen, C.J.Siebert, M.-D.Zhu, J.Chromatogr., **403** (1987) 47.
Carrier-free zone electrophoresis, displacement electrophoresis and isoelectric focusing in a high-performance electrophoresis apparatus.

[10] B.Nickerson, J.W.Jorgenson, J.High Resolut.Chromatogr. Chromatogr. Commun., **11** (1988) 533.
High speed capillary zone electrophoresis with laser induced fluorescence detection.

[11] F.Foret, M.Deml, P.Boček, J.Chromatogr., **452** (1988) 601.
Capillary zone electrophoresis. Quantitative study of the effects of some dispersive processes on the separation efficiency.

[12] S.Hjertén, Electrophoresis (Weinheim), **11** (1990) 665.
Zone broadening in electrophoresis with special reference to high-performance electrophoresis in capillaries : an interplay between theory and practice.

[13] J.Vindevogel, P.Sandra, Anal.Chem., **63** (1991) 1530.
Resolution optimization in micellar electrokinetic chromatography : Use of Plackett-Burman statistical design for the analysis of testosterone esters.

[14] J.Vindevogel, P.Sandra, J.Chromatogr., **541** (1991) 483.
Simultaneous pH and ionic strength effects and buffer selection in capillary electrophoretic techniques.

[15] S.Fujiwara, S.Honda, Anal.Chem., **59** (1987) 2773.
Determination of ingredients of antipyretic analgesic preparations by micellar electrokinetic capillary chromatography.

[16] R.K.Momii, N.H.Nachtreib, Inorg.Chem., **6** (1967) 1189.
Nuclear magnetic resonance study of borate-polyborate equilibria in aqueous solution.

[17] P.K.Dasgupta, O.Nara, Anal.Chem., **62** (1990) 1117.
Measurement of acid dissociation constants of weak acids by cation exchange and conductometry.

4 Capillary Electrochromatography

4.1 From Electrophoresis to Electrochromatography

Electrophoresis and chromatography are based on different separation principles. Some collision exists, on exploiting sorption in electrophoresis, as well as on using electromotion in chromatography. With the advent of capillary electrophoresis both techniques became firmly integrated. As will be shown, Micellar Electrokinetic Chromatography (MEKC) has been most successful, in number as well as in variety of applications. Other attempts at combining electrophoresis and chromatography will be briefly discussed here.

In classical electrophoresis, the term *electrochromatography* has been used when electrophoretic mobilities are influenced by sorption to the gel [1] but equally for two-stage modes (SDS-PAGE, followed by gel filtration) [2]. It has even been used for that form of chromatography where the sorptive properties of the stationary phase are influenced by applying a small potential [3], a technique better described as electrochemical chromatography [4]. *Knox* [5] has used the name electrochromatography for all separation methods in which the flow results from electroosmosis, including MEKC and even CZE, although in the latter sorption is, in principle, absent and electroosmosis is not essential. We prefer to limit *electrochromatography* and *capillary electrochromatography (CEC)* to those methods were stationary phases typical for liquid chromatography are used in combination with electroosmosis, as distinct from CZE but also from MEKC.

4.2 Capillary Electrochromatography (CEC)

4.2.1 Electroosmotic Propulsion in Chromatography

As long ago as 1954, *Mould* and *Synge* used electroosmotically induced flow over porous membranes that would not have resisted pressure-induced stresses [6]. This technique, termed 'electrokinetic ultrafiltration' was applied to separate polysaccharides on collodion membranes. The awareness that the flow profiles for pressure-induced and high-voltage induced electroosmotic flow were different (**Figure 4.1**), and this would have consequences for peak dispersion, led to the application in column chromatography [7].

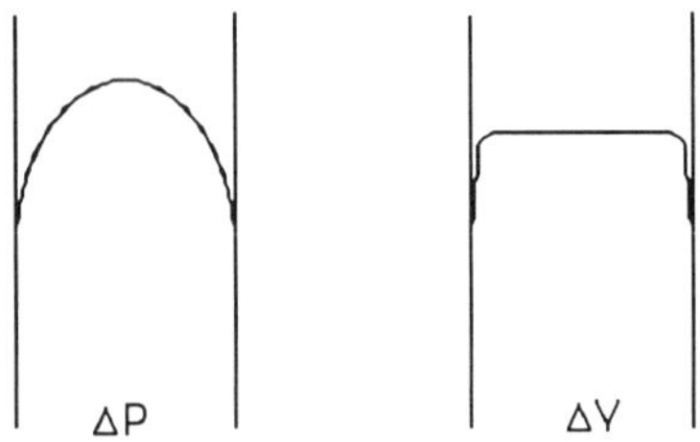

Figure 4.1
Pressure and voltage driven flow, parabolic versus flat flow profile.

The effect of the flow profile on plate height is most obvious in the case of open columns (packed columns have been discussed by *Knox* [8]). In the case of flow through open tubes, the plate height is determined by the Golay equation (ignoring mass transfer in the stationary phase) [9] :

$$h = \frac{2\,D_m}{v} + \frac{C_m\,r_c^2\,v}{D_m} \tag{4.1}$$

In pressure-driven chromatography, using a laminar flow with a parabolic profile, the resistance to mass transfer in the mobile phase, C_m, is related to the capacity factor k' by :

$$C_m = \frac{11\,k'^2 + 6\,k' + 1}{24\,(1 + k')^2} \tag{4.2}$$

Miniaturization of the column radius (eq.(4.1)) is one method to reduce the contribution of the flow profile to the plate height. Another is the elimination of the parabolic flow profile by using electroosmotic flow instead of pressure-driven flow. For a perfectly flat profile, C_m is given by :

$$C_m = \frac{k'}{4\,(1 + k')^2} \tag{4.3}$$

Values of C_m for parabolic and flat flow profiles are compared in **Figure 4.2**. More accurate calculations [10], considering a more realistic electroosmotic flow profile

(Section 2.2.2), show that deviations from eq.(4.3) are less than 5% for $k' \geq 1$.

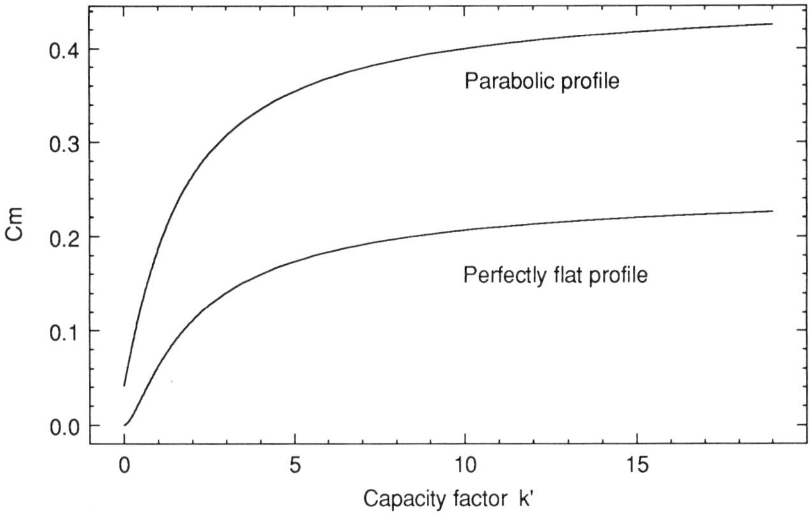

Figure 4.2
C_m as a function of the capacity factor for different flow profiles.

4.2.2 Packed Columns

Pretorius was the first to apply high voltages over columns (1 mm ID) packed with a microparticulate chromatographic phase (75-125μm Partisil) [7]. He demonstrated that adequate ζ_{eo} potentials could be developed using typical reversed-phase eluents ($MeOH:H_2O$) on octane-coated silica gel. Although the reduced plate height could be demonstrated with an unretained solute, the absence of suitable injection and detection systems prevented analytical application. The development of CZE by *Everaerts* and *Jorgenson* (1979-1981) caused a revival in the interest in electroosmotically driven chromatography.

Jorgenson [11], having solved instrumental problems, showed the separation of 9-methylanthracene and perylene on a 170 μm ID column packed with 10 μm octadecylated silicagel (ODS), with acetonitrile as the mobile phase and using an on-column fluorescence detector (**Figure 4.3**). Similarly, *Knox* [5] reported the separation of polyaromatic hydrocarbons in a 50μm ID column with 5 μm ODS-packing. Efficiencies ranged from 30,000 [11] to 60,000 [5] theoretical plates. *Tsuda* [12,13] combined pressure and voltage to separate compounds with similar capacity factors but different charges. This technique was eventually modified into an on-column sample enrichment system [14].

The effect of the particle diameter on the flow profile and the influence of the mobile phase composition on electroosmosis have been studied [5,15]. The contributions from the packing particles to the plate height have been discussed and it has been suggested that, using sub-micron particles, all other contributions than axial diffusion should become insignificant [8]. This has, however, not yet been realized.

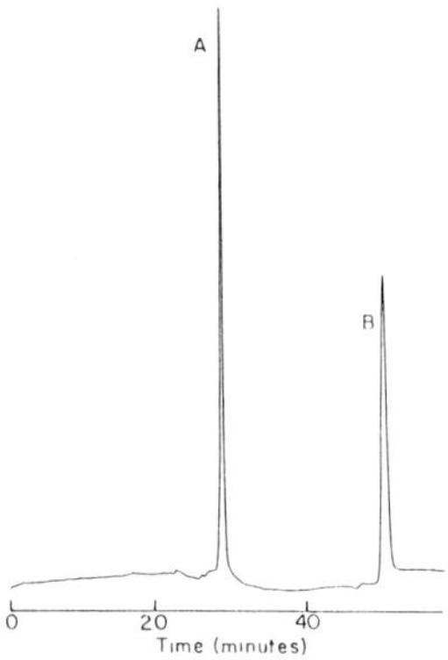

Figure 4.3
Electrochromatographic separation of 9-methylanthracene (A) and perylene (B) in a packed capillary column. *(Reprinted with permission [11], © Elsevier Science Publishers B.V.)*

4.2.3 Open Columns

From eq.(4.1), it follows that, in open-tubular liquid chromatography, very small column diameters are required to obtain high efficiencies. The same holds in electrochromatography. Using 30 μm ID, ODS-wall-coated columns, *Tsuda* could only obtain 10,000 theoretical plates [16]. Reducing the internal diameter however requires also the use of more sophisticated detectors. Using laser-induced fluorescence, efficient separations of polycyclic aromatics were realized by *Bruin* et al. (10 μm ID, ODS wall-coated) [17] (**Figure 4.4**) and by *Pfeffer* and *Yeung* (10 μm

ID, siloxane wall-coated) [18]. In the latter case, a cationic surfactant was used at a concentration below the critical micelle concentration to enhance the electroosmotic flow.

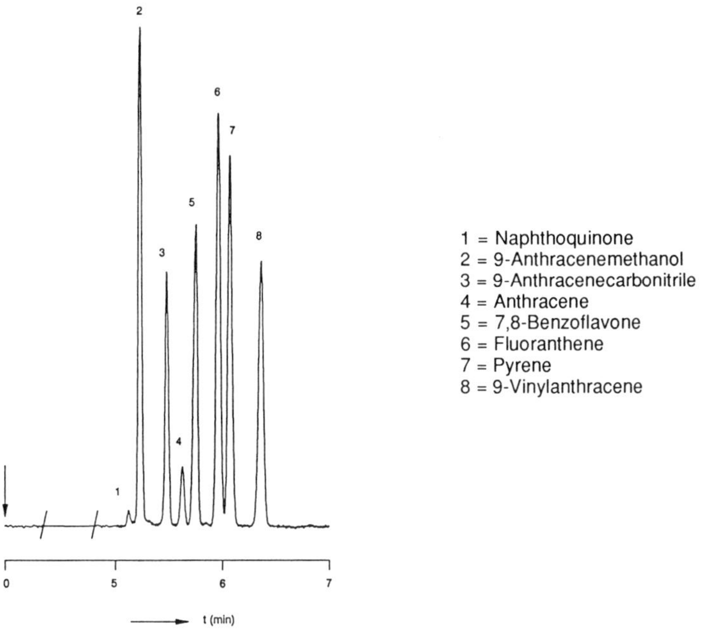

Figure 4.4
Electrochromatographic separation of polycyclic aromatic hydrocarbons in an open, wall-coated capillary column. *(Reprinted with permission [17], © Elsevier Science Publishers B.V.)*

4.2.4 The Future of Capillary Electrochromatography

Some chromatographers have expected much from using electroosmosis in combination with conventional stationary phases. This optimism is sharply contrasted with the limited number of separations that have actually been reported. Very probably, experimental problems can explain this discrepancy. Right at the beginning, *Jorgenson* remarked [11] that CEC in packed columns is difficult and less convenient in controlling mobile phase flow, expressing doubts whether the performance increase would justify increased difficulty. Packed 320 μm ID columns are routinely applied in micro-LC, but further reduction of the internal diameter makes packing much more difficult. Nevertheless, it has been predicted [20] that packed columns would be more robust in relation to electroosmotic flow variability, and that a gain in popularity has to be expected.

Although open-tubular CEC may be simpler than its pressure-driven counterpart (no high pressure-resisting fittings needed, valveless injection[1], less restrictive in optimal diameter) [18], the small column volume demands detector requirements that are not routinely available. More seriously, calculations have shown that the improved efficiency, expected from the electroosmotic flow profile, will only be realized at small capacity factor values [10,20]. This is also suggested by **Figure 4.5a**. As k' increases, loss of efficiency due to increased mobile phase velocity becomes more apparent, preventing fast analysis. Similar calculations for proteins (low D_m, **Figure 4.5b**) confirm the superiority of zone electrophoresis (k' = 0) and the need to avoid chromatographic effects resulting from interaction with the column wall.

1 *The same injection techniques that will be discussed in Part II are used.*

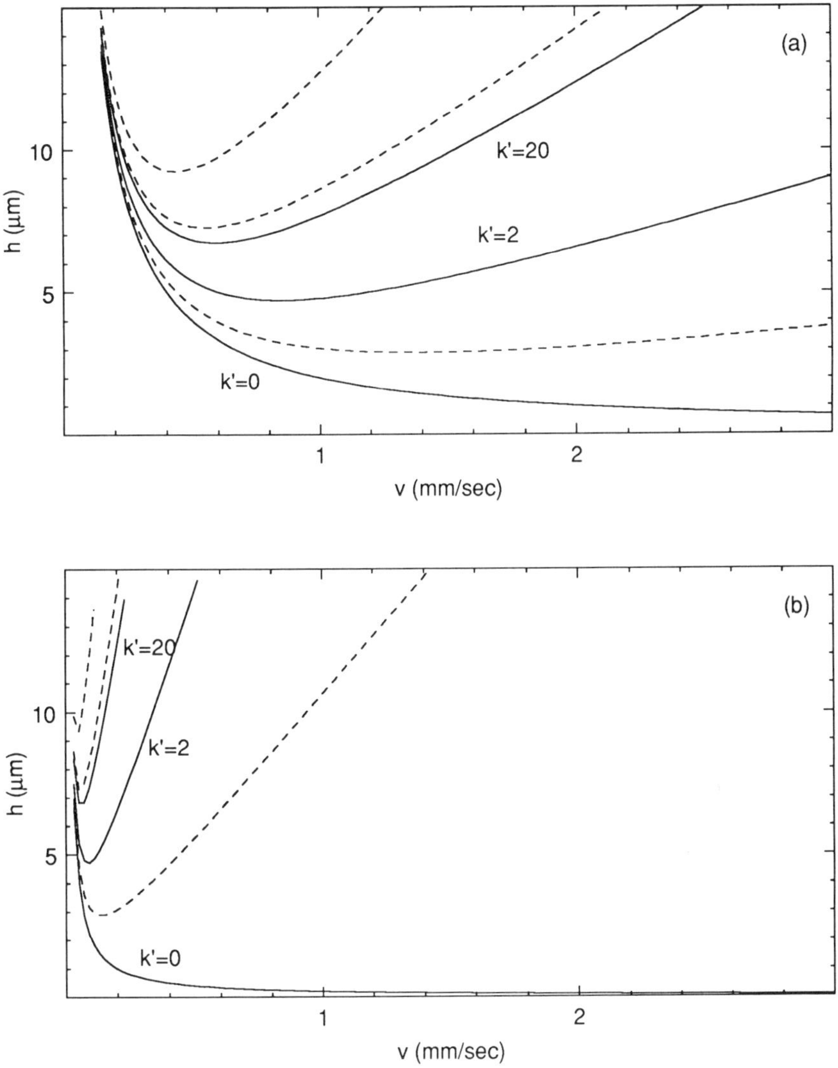

Figure 4.5

Plate height as a function of linear velocity in a 10 μm ID capillary, according to eq.(4.1). Solid lines : perfectly flat profile. Broken lines : parabolic flow profile.
(a) Small molecules ($D_m = 10^{-9}$ m^2/sec)
(b) Large molecules ($D_m = 10^{-10}$ m^2/sec)

References

[1] J.S.Krauss, M.H.Jonah, Clin.Chem., **28** (1982) 2000.
Citrate agar electrophoresis is electrochromatography.

[2] S.Otsuka, L.Listowsky, Anal.Biochem., **102** (1980) 419.
High-resolution preparative electrochromatography for purification of two subunit types of ferritin.

[3] R.F.Antrim, R.A.Scherrer, A.M.Yacynych, Anal.Chim.Acta, **164** (1984) 283.
Electrochromatography - A preliminary study of the effect of applied potential on a carbonaceous chromatographic column.

[4] H.Ge, P.R.Teasdale, G.G.Wallace, J.Chromatogr., **544** (1991) 305.
Electrochemical chromatography - packings, hardware and mechanisms of interaction.

[5] J.H.Knox, I.W.Grant, Chromatographia, **24** (1987) 135.
Miniaturisation in pressure and electroendosmotically driven liquid chromatography : Some theoretical considerations.

[6] D.L.Mould, R.L.Synge, Biochem.J., 58 (1954) 571.
Separations of polysaccharides related to starch by electrokinetic ultrafiltration in collodion membranes.

[7] V.Pretorius, B.J.Hopkins, J.D.Schieke, J.Chromatogr., **99** (1974) 23.
Electro-osmosis. A new concept for high-speed liquid chromatography.

[8] J.H.Knox, Chromatographia, **26** (1988) 329.
Thermal effects and band spreading in capillary electro-separation.

[9] J.C.Giddings, "Dynamics of Chromatography. Part I : Principles and Theory", Dekker Inc, New York, 1965.

[10] M.Martin, G.Guiochon, Anal.Chem., **56** (1984) 614.
Axial dispersion in open-tubular capillary liquid chromatography with electroosmotic flow.

[11] J.W.Jorgenson, K.D.Lukacs, J.Chromatogr., **218** (1981) 209.
High-resolution separations based on electrophoresis and electroosmosis.

[12] T.Tsuda, Anal.Chem., **59** (1987) 521.
Electrochromatography using high applied voltage.

[13] T.Tsuda, Anal.Chem., **60** (1988) 1677.
Chromatographic behavior in electrochromatography.

[14] T.Tsuda, Y.Muramatsu, J.Chromatogr., **515** (1990) 645.
Electrochromatography with continuous sample introduction.

[15] T.S.Stevens, H.J.Cortes, Anal.Chem., **55** (1983) 1365.
Electroosmotic propulsion of eluent through silica based chromatographic media.

[16] T.Tsuda, K.Nomura, G.Nakagawa, J.Chromatogr., **248** (1982) 241.
Open-tubular microcapillary liquid chromatography with electro-osmosis flow using a UV detector.

[17] G.J.M.Bruin, P.P.H.Tock, J.C.Kraak, H.Poppe, J.Chromatogr., **517** (1990) 557.
Electrically driven open-tubular liquid chromatography.

[18] W.D.Pfeffer, E.S.Yeung, Anal.Chem., **62** (1990) 2178.
Open-tubular liquid chromatography with surfactant enhanced electroosmotic flow.

[19] J.H.Knox, K.A.McCormack, J.Liq.Chromatogr, **12** (1989) 2435.
Capillary electroseparation methods : looking into the crystal ball.

[20] M.Martin, G.Guiochon, Y.Walbroehl, J.W.Jorgenson, Anal.Chem., **57** (1985) 559.
Peak broadening in open-tubular liquid chromatography with electroosmotic flow.

5 Micellar Electrokinetic Chromatography (MEKC)

5.1 Separation Mechanism

5.1.1 Neutral Solutes

Micellar Electrokinetic Chromatography (MEKC) was developed by *Terabe* [1,2], after a suggestion by *Nakagawa* [3,4], for the electrokinetic chromatography of neutral molecules. Others have preferred the abbreviation MECC (Micellar Electrokinetic Capillary Chromatography) [5]. Following *Otsuka* and *Terabe* [6], we will continue to use MEKC to stress the electrokinetic nature of the separation, which already implies the capillary format.

Using the same instrumentation as CZE, MEKC requires only the addition of a charged surfactant to the background electrolyte. When the surfactant concentration is larger than the critical micelle concentration (CMC), micelles are formed. The situation shown in **Figure 5.1** refers to the analysis of neutral molecules with an anionic surfactant and a neutral or alkaline buffer in fused silica capillaries. In this case, which is the most common situation, the electroosmotic flow μ_{eo} is directed to the cathode. The micelles are negatively charged and, having their own electrophoretic mobility $\mu_{ep,MC}$, they will move with a mobility $\mu_{MC} = \mu_{eo} - \mu_{ep,MC}$. When the electroosmotic flow is high enough, the micelles will move in the same direction as the bulk of the buffer, but at a slower velocity.

Analyte molecules will be distributed between the aqueous and micellar phases. Very polar molecules that occur only in the aqueous phase will elute with a mobility μ_{eo} and be detected at t_0. This is the shortest elution time possible in the system and can, by analogy, be compared to the void volume concept in column chromatography. Highly hydrophobic compounds, occurring in the micelle interior all the time, will elute at the corresponding t_{MC}. All the analytes must elute between those two limits, t_0 and t_{MC}, depending upon their partition between the aqueous and the micellar phases, as expressed by the retention ratio

$$R_r = \frac{n_{aq}}{n_{aq} + n_{MC}} = \frac{1}{1 + \widetilde{k}'} \qquad (5.1)$$

The effective linear velocity can be expressed as

$$v_R = R_r\, v_{eo} - (1 - R_r)\, v_{MC} \qquad (5.2)$$

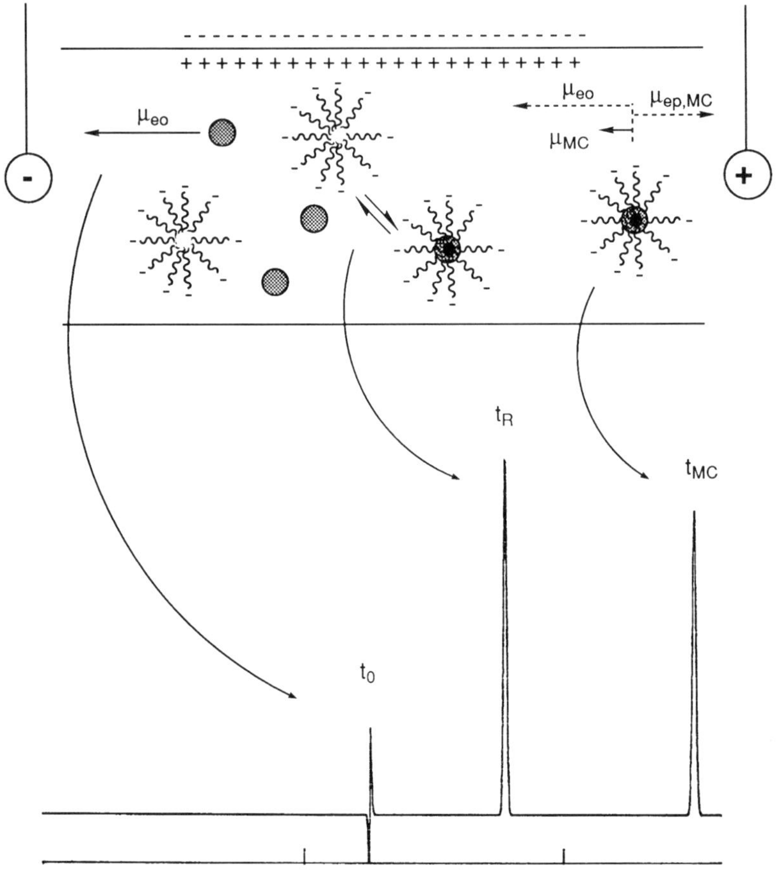

Figure 5.1
Separation mechanism in Micellar Electrokinetic Chromatography.

or after combination with eq.(5.1), and using the inverse relationship between t_R and v_R [2]

$$t_R = \frac{1 + \widetilde{k}'}{1 + \frac{t_0}{t_{MC}} \widetilde{k}'} \cdot t_0 \tag{5.3}$$

and also by inversion, defining the capacity factor $\tilde{k}'$ [2]

$$\tilde{k}' = \frac{t_R - t_0}{t_0\,(1 - \frac{t_R}{t_{MC}})} \tag{5.4}$$

Equations (5.3) and (5.4) look slightly more complicated than their counterparts in chromatography [1]. The differences are caused by the fact that the phase which functionally corresponds to the stationary phase is actually moving. Therefore it is called a 'pseudo-stationary phase'. If the micellar phase is really stationary, a condition which can be realized by manipulating the electroosmotic flow so that $\mu_{eo} = -\mu_{ep,MC}$ (equal but opposite), then t_{MC} becomes infinite, simplifying eq.(5.3) and (5.4) to the familiar equations :

$$t_R = (1 + \tilde{k}').\ t_0 \tag{5.5}$$

and

$$\tilde{k}' = \frac{t_R - t_0}{t_0} \tag{5.6}$$

The consequences of the moving pseudo-stationary phase are obvious. In isocratic chromatography, a series of peaks with constant k' increment will elute at evenly spaced time intervals, according to eq.(5.5). But in MEKC, the time interval between peaks will be reduced as $\tilde{k}'$ increases, resulting in peak compression near t_{MC} (**Figure 5.2-5.3**). In this respect, the analogy between MEKC and gradient chromatography has been observed [2], although in the latter case the compression follows from manipulation of the mobile phase.

Calculation of capacity factors requires knowledge of t_0 and t_{MC}. As in CZE, t_0 is determined by measuring the migration time of small, neutral molecules, sufficiently polar to avoid partition with the micelles. The UV absorbance of formamide [7,8] or acetone [9,10] but also the baseline disturbance caused by methanol [2,7,8] can be used.

Following a practice in tracer electrophoresis [11,12], t_{MC} is measured by adding a water-insoluble dye to the sample. Being completely included in the micelle, the migration behavior of the dye is considered to be representative for the micelles. Sudan III [2] is the most common t_{MC} marker, but also Sudan IV [13,14],

---

[1] *The symbol $\tilde{k}'$ is used instead of k' to emphasize the difference.*

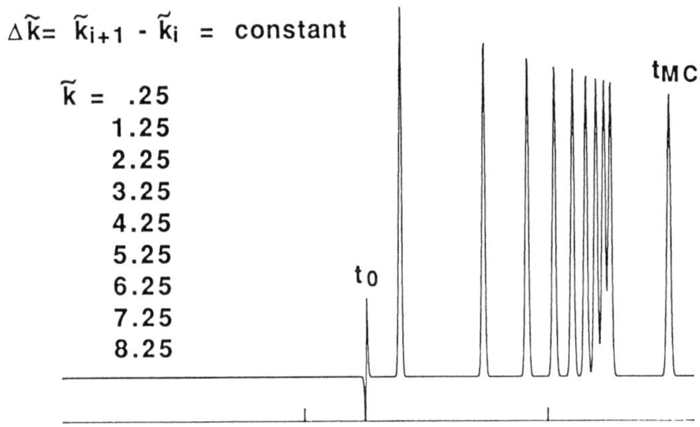

Figure 5.2
Relation between retention time and capacity factor (simulated chromatogram).

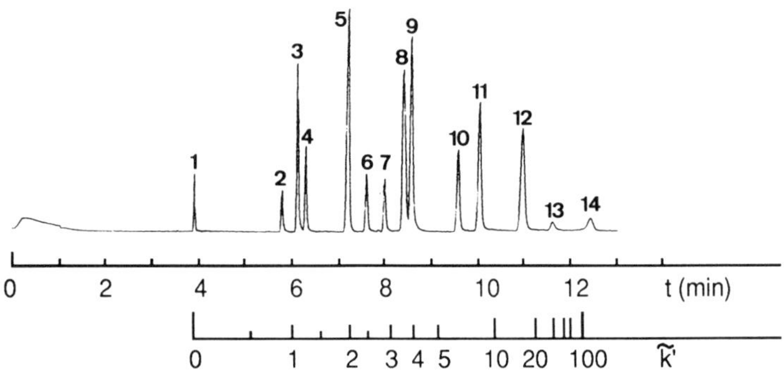

Figure 5.3
MEKC separation of aromatic compounds. Buffer : 110 mM SDS in 25 mM borate, pH 8.5. Column : 75 μm x 50/57 cm. Applied voltage : 20 kV. Detection : UV, 200 nm. 1 = Formamide (t_0), 2 = Aniline, 3 = Phenol, 4 = Benzyl alcohol, 5 = Benzoic acid, 6 = Benzaldehyde, 7 = Nitrobenzene, 8 = Phenylacetone + Benzyl cyanide, 9 = Acetophenone, 10 = Toluene, 11 = Chlorobenzene, 12 = Ethylbenzene + o-Xylene, 13 = Naphthalene, 14 = Sudan III (t_{MC}).

Yellow OB [9,15,16], Timepidium bromide [14,17] and Orange OT [18] have been used (**Figure 5.4**). Aromatic hydrocarbons like anthracene or 9-methyl anthracene are not reliable as t_{MC} markers [7,8].

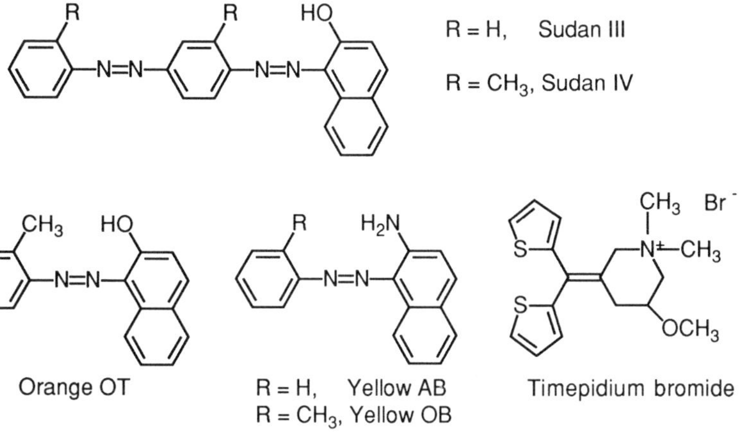

Figure 5.4
Chemical structure of some t_{MC} markers.

No suitable t_0 marker is known that can be used with fluorescence detection [19]. Coumarin dyes [19,20], NBD-dodecylamine [21] and nitropyrene [22] have been used for the determination of t_{MC} with fluorescence detection.

Especially when organic cosolvents are used, doubts have been expressed about the complete micellization upon which the use of micelle tracers is based [7,8,19,21]. In that case, an iterative procedure has been proposed for a series of homologous compounds to check the validity of t_{MC} [7,8]. Using the highest retention time as a t_{MC} seed value, $\tilde{k}'$ values are calculated (**Table 5.1**). From the correlation between $\log(\tilde{k}')$ and the carbon number (**Figure 5.5**), a new $\tilde{k}'$-value is calculated that is used to estimate a new t_{MC} according to eq.(5.3) and the known t_R and t_0. This procedure is repeated using the last obtained t_{MC} value until the difference between consecutive t_{MC} or $\tilde{k}'$ values is considered negligible. The linearity between $\log(\tilde{k}')$ and carbon number has also been used to certify $\tilde{k}'$ values, based on t_{MC} values obtained in the conventional way [13,23].

Table 5.1 Iterative calculation of t_{MC} [7,8] from the migration time of a series of homologous dansyl amines (based on data from [7], t_0 = 1625 sec). Buffer : 25 mM SDS in a phosphate/borate buffer, pH 8.0, containing 10% methanol. Column : 50 µm x 105/120 cm. Applied voltage : 30 kV.

Compound a)	i = 1 b)	i = 2	i = 3	i = 4	i = 5	...	i = 13	i = 14
C_4 (10903)	12.8	12.2	11.9	11.8	11.7	...	11.5	11.5
C_5 (14800)	32.7	29.3	27.7	26.9	26.4	...	25.8	25.8
C_6 (17778)	103.2	75.5	66.1	61.7	59.4	...	56.3	56.3
C_7 (19672)	283.0	185.3	154.9	141.3	134.2	...	124.8	124.7
t_{MC} (assumed)	19672	20475	20927	21192	21351	...	21592	21594
t_{MC} (calculated)	20475	20927	21192	21351	21447	...	21594	21595 c)

a) $C_{4\text{-}7}$ are dansylated butylamine, pentylamine, hexylamine and heptylamine respectively. The number in parenthesis is the migration time in seconds.
b) Numbers indicate capacity factors for $C_{4\text{-}7}$ and migration times in seconds for t_{MC}.
c) In [7] only the final result, 21326 sec, is mentioned. Probably, the discrepancy is not just due to rounding errors but to a lower number of iterations.

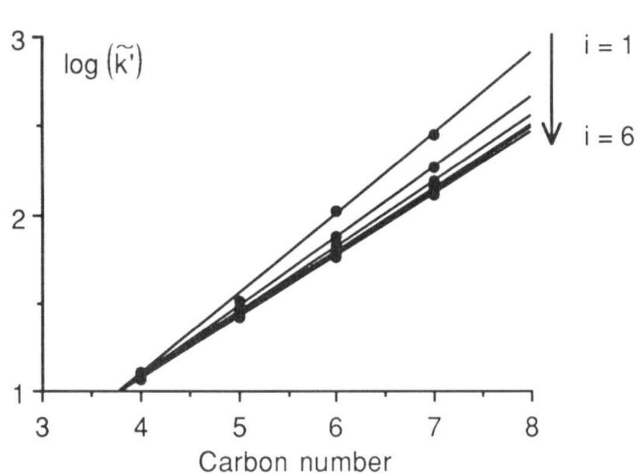

Figure 5.5

Log($\tilde{k}'$) vs. carbon number relationship used in the iterative calculation of $\tilde{k}'$ and t_{MC} for the data in **Table 5.1** (i = iteration number).

5.1.2 Charged Solutes.

MEKC was originally conceived for the electrokinetic analysis of neutral molecules. It can however equally well be applied to the analysis of ionic compounds. In some of these cases, the selection of MEKC, as opposed to CZE, logically follows from the fact that both charged and uncharged compounds occur in the same mixture (e.g. in the mixture of **Figure 5.3**, benzoic acid is completely ionized at the pH of the buffer). In other cases the stereoselectivity, obtained by using chiral micelles, is exploited. For some specific applications, the superior resolving power of MEKC, as compared to CZE, has been acknowledged. The partition between the aqueous and micellar phases increases the selectivity and allows the separation of ions with very similar electrophoretic mobilities. As an example, the analysis of hop bitter acids will be briefly discussed.

The hop acids, especially the β-acids, are only soluble in aqueous media when fully dissociated. In an alkaline buffer and with a strong electroosmotic flow, it was possible to let the anions migrate to the cathode (**Figure 5.6**) but separation is unsatisfactory. The observed partial peak splitting can be ascribed to the co-homologues, which are expected to have a slightly higher electrophoretic mobility. It has already been demonstrated that increasing mobility differences is the most effective way to improve resolution (Chapter 3). Here, this would require pH control to differentiate between the normal and the ad-homologues which have very similar pK_a values [24] which is impossible for solubility reasons. This justifies the use of surfactants to obtain resolution of the ionic solutes (**Figure 5.7**). The β-acids elute before the α-acids although the latter are more polar [25]. Indeed, the migration behaviour is controlled by the degree of micellar solubilization but also by the ion's own electrophoretic mobility in the aqueous phase. Apparently, the effective electrophoretic mobility, being lowest for the β-acids, determines group separation, which is then modulated by polarity differences between group members. Within each group the C_3-side chain co-homologue elutes before the C_4-side chain ad- and normal homologues.

Although the influence of the micellized ion on the micelle mobility can be ignored, the effect of the ion's own mobility in the aqueous phase is important. It also makes the determination of capacity factors more difficult. By analogy with eq.(5.2) [15], and with v_{ion} representing the apparent electrophoretic mobility of the ion (acc. to eq.(2.18)) in the aqueous phase of the micellar medium :

$$v_R = R_r\, v_{ion} - (1 - R_r)\, v_{MC} \tag{5.7}$$

which leads to (with t_{ion} the hypothetical migration time of the ion in the aqueous phase, but unaffected by partition with the surfactant !) :

$$\tilde{k}' = \frac{t_R - t_{ion}}{t_{ion}\left(1 - \frac{t_R}{t_{MC}}\right)} \tag{5.8}$$

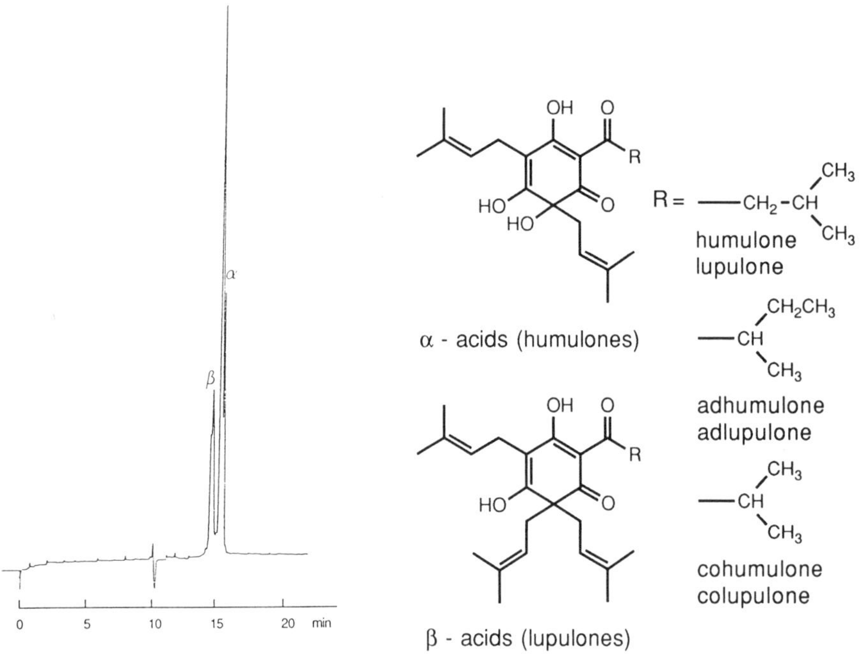

Figure 5.6
CZE analysis of hop bitter acids in a hop extract. Buffer : 50 mM TRIS/HOAc, pH 8.6. Column : 100 µm x 50/80 cm. Applied voltage : 12 kV. Detection : UV, 220 nm. *(Reprinted with permission [25], © Hüthig Verlag)*

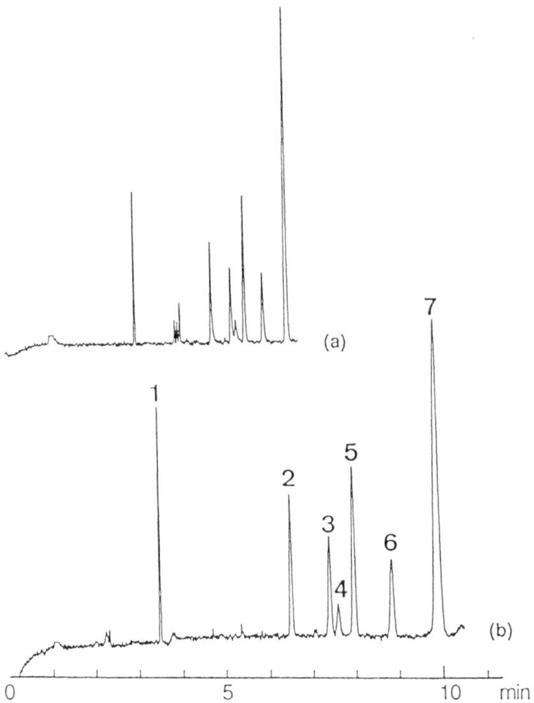

Figure 5.7
MEKC analysis of hop bitter acids in a hop extract. Buffer : 40 mM SDS in 25 mM borate (a) or TRIS/HCl (b), pH 8.55. Column : 75 μm x 43/50 cm. Applied voltage 20 kV (at 20°C). Detection : UV, 214 nm. 1 = formamide (t_0 marker), 2 = colupulone, 3 = lupulone, 4 = adlupulone, 5 = cohumulone, 6 = adhumulone, 7 = humulone.
(Reprinted with permission [24], © Hüthig Verlag)

or as it is usually formulated [15]

$$\tilde{k}' = \frac{v_R - v_{ion}}{v_{MC} - v_R} = \frac{v_R^* - v_{ep,ion}}{v_{ep,MC} - v_R^*} \qquad (5.9)$$

where v_{ion} = linear velocity of the ion in the aqueous phase
$v_{ep,ion}$ = $v_{eo} - v_{ion}$
v_R = observed linear velocity
v_R^* = $v_{eo} - v_R$

The calculation of the capacity factors requires knowledge of t_{ion} or $v_{ep,ion}$, quantities which are not readily accessible. Some authors have based their calculations on the observed mobilities in the absence of the surfactant [18,26] but it remains to be demonstrated how the resulting $\tilde{k}'$ values are biased. From eq.(5.7), the following mobility relation can be derived (after reducing velocities to mobilities) :

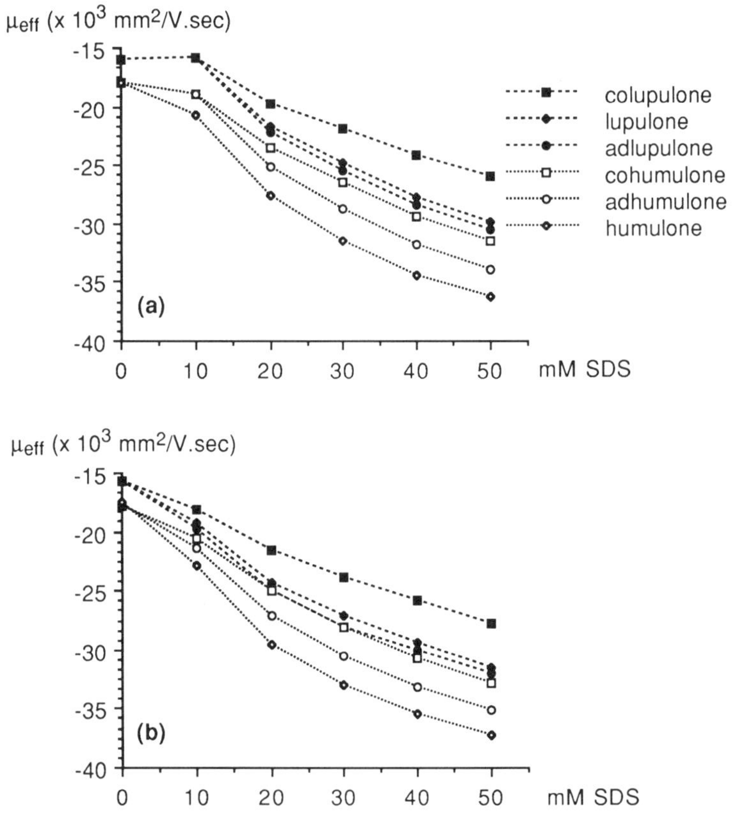

Figure 5.8
Effective mobility as a function of surfactant concentration. (a) Borate buffer. (b) TRIS buffer. Conditions as in **Figure 5.7**. *(Reprinted with permission [24], © Hüthig Verlag)*

$$\mu_{eff} = \frac{1}{1+\tilde{k}'}\mu_{ep,ion} + \frac{\tilde{k}'}{1+\tilde{k}'}\mu_{ep,MC} \qquad (5.10)$$

The migration times in the two chromatograms of **Figure 5.7** are different which is related to differences in the electroosmotic flow. Nevertheless the effective mobilities (expressed as negative numbers because they are opposed to the electroosmotic mobility) are fairly similar in the two buffer systems (**Figure 5.8**). Extrapolation to zero surfactant concentration, which is equivalent to $\tilde{k}' = 0$ in eq.(5.10), leads to values which differ considerably from the ones observed in the absence of the surfactant (**Table 5.2**).

Table 5.2 Observed and extrapolated values for effective electrophoretic mobility (x 10^3 $mm^2/V.sec$, conditions as in **Figure 5.8**) *(Reprinted with permission [24], © Hüthig Verlag)*

	Borate	Borate/SDS[a)]	TRIS	TRIS/SDS[a)]
Colupulone	-16.0	-10.3	-15.7	-13.1
Adlupulone	-16.0	-6.1	-15.7	-11.3
Lupulone	-16.0	-6.9	-15.7	-11.7
Cohumulone	-17.9	-13.0	-17.7	-14.2
Adhumulone	-17.9	-9.9	-17.4	-12.7
Humulone	-17.9	-10.7	-17.4	-12.4

a) Extrapolated to $c_{SDS} = 0$

5.2 Surfactants

5.2.1 Anionic Surfactants

5.2.1.1 Alkyl Sulfates and Carboxylates

As the surfactant must be charged, sulfates and sulfonates are generally (but not exclusively) preferred over carboxylates. Sodium dodecyl sulfate (SDS, sodium lauryl sulfate) is the most frequently used surfactant. In MEKC, it has attained the same status as octadecylated silica gel in reversed-phase liquid chromatography. The CMC of SDS is approximately 8 mM in pure water (**Table 5.3**) and this figure is still often cited although it was established a long time ago (e.g.[12]) that this value is is lowered by the presence of electrolytes. *Saitoh* et al. have measured the CMC of SDS in typical MEKC buffer solutions and found values ranging from 2.6 to 5.0 mM [9].

Homologues of SDS seems less suitable as surfactants in MEKC. Sodium decyl sulfate has been reported to cause excessive currents at concentrations above 50 mM, resulting in Joule heating problems [27,28]. Sodium tetradecyl sulfate has low solubility, restricting its use at elevated temperatures (≥ 35 °C, [2,29]). Other anionic surfactants, like sodium dodecyl sulfonate [2] and sodium lauroylmethyl taurate [30,31] have occasionally been used. Long-chain N-substituted L-alaninates and L-valinates have been applied in combination with SDS [32,33] or pure [13,34] to impart chiral selectivity to the micelles.

Table 5.3 Critical micelle concentrations and aggregation numbers of some surfactants. Data refer to aqueous solutions at ambient temperature. In buffer solutions the CMC is reduced and n_{aggr} is increased [a)].

Surfactant	CMC (mM)	n_{aggr}	
Sodium decyl sulfate	33	41	[35]
Sodium dodecyl sulfate	8.2	64	[35]
Sodium tetradecyl sulfate	2.05	80	[35]
Sodium hexadecyl sulfate	0.45	100	[35]
Sodium lauroylmethyl taurate	8.7	-	[31]
Sodium cholate	13	-	[36]
Sodium dehydrocholate	10	-	[36]
Sodium taurocholate	10	-	[36]
Sodium taurodehydrocholate	6	-	[36]
Decyltrimethylammonium bromide	65	-	[37]
Dodecyltrimethylammonium bromide	15	50	[38]
Tetradecyltrimethylammonium bromide	3.5	75	[38]
Hexadecyltrimethylammonium bromide	0.92	61	[38]

a) For a discussion of these effects, see [37].

5.2.1.2 Bile Salts

Unconjugated and taurine-conjugated bile salts (**Figure 5.9**) form aggregates in aqueous solution but, unlike typical anionic detergents, they have relatively small aggregation numbers (< 10) with a less sharply defined CMC [36,37]. Bile salt micelles[1] have been used as pseudo-stationary phases in MEKC for 2 reasons. The solubilizing power is lower than with SDS, allowing the analysis of hydrophobic compounds [22,23,39] and the chiral properties have been used for enantiomeric resolution [22,26,40,41].

The use of unconjugated bile salts is restricted to neutral or alkaline buffers (pH 7-9) but taurine-conjugated bile salts can also be applied in more acidic buffers (pH 3) [26]. STLC cannot be used, due to gelation [26] and SDHC showed no separation power [23,39] in accordance with the fact that no CMC could be measured for SDHC or STDHC [36]. The relative performance of SC, STC, SDC and STDC is determined by the better solubilizing power of the deoxy-forms in relation to the structure of the analyte [23,40,41].

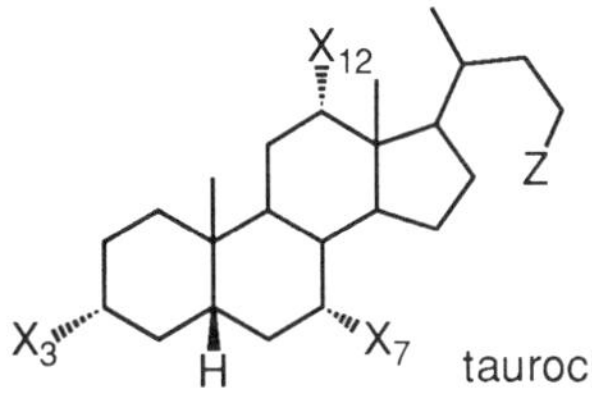

cholates

$Z = COO^- Na^+$

taurocholates

$Z = CONHCH_2CH_2SO_3^- Na^+$

	X_3	X_7	X_{12}		X_3	X_7	X_{12}
SC	OH	OH	OH	STC	OH	OH	OH
SDC	OH	H	OH	STDC	OH	H	OH
SDHC	=O	=O	=O	STDHC	=O	=O	=O
				STLC	OH	H	H

Figure 5.9
Bile salts and their tauro-conjugates.
S(T)C : sodium (tauro)cholate
S(T)DC : sodium (tauro)deoxycholate
S(T)DHC : sodium (tauro)dehydrocholate
STLC : sodium taurolithocholate

5.2.2 Cationic Surfactants

The principles of MEKC separation with cationic surfactants [16], almost always hydrophobic-chain ammonium salts, are very similar although the charge on

[1] *Some consider bile salts as nonmicelle forming amphiphiles [37] and prefer to use the term 'multimers' [36].*

the micelles is reversed. Even at low concentrations, below the critical micelle concentration, the electroosmotic flow is reversed by the presence of the surfactant, preserving the normal relation between the direction of the bulk flow and the micelle migration. **Figure 5.1** can equally well be used to describe MEKC with cationic surfactants, provided that all plus signs are replaced by minus signs, and vice versa, including the reversal of the applied electric field.

The dependence of the electroosmotic flow on surfactant concentration has been determined [42-46] and cationic surfactants have been used at concentrations below the CMC to reverse the electroosmotic flow in capillary zone electrophoresis [47-49]. In MEKC, cationic surfactants have been used for the analysis of PTH-amino acids [16], aromatic compounds [27], nucleotides [43,50] (**Figure 5.10**) and derivatized oligosaccharides [45].

5.3 Relation Between Micellar Electrokinetic Chromatography and Electrophoresis

In the simplest case, CZE separation is based on intrinsic mobility differences between ions. When the differences are absent or too small to obtain separation, additional equilibria are exploited to influence the charge/mass-ratio. **Table 5.4** summarizes a number of methodologies according to 4 criteria : the role of intrinsic mobility differences ($\Delta\mu_{intr}$), the role of equilibria that affect the charge of the analyte, the role of equilibria that affect the mass of the analyte and the degree of phase separation. Pure electrophoresis and chromatography can be found at the top and bottom of the table respectively.

Cases 1 to 5 will be accepted as forms of electrophoresis without much discussion, although buffers containing soluble chiral Cu^{2+}-amino acid-complexes have been labeled as pseudophases [51]. Case 4 is exploited in the electrophoresis of diols as borate-complexes [52] or for the analysis by solvophobic association in non-aqueous medium [53].

Cases 14 and 15 are easily recognized as chromatography. They possess the characteristic properties of phase separation and lateral (or radial) displacement [54]. Case 15 (neutral analytes combined with a true stationary phase) is the only case that relies essentially on electroosmotic flow for displacement.

Cases 12 and 13 are now generally accepted as forms of (electrokinetic) chromatography although phase separation is only apparent on a microscale and the importance of lateral transport is much reduced by the dispersion of one phase

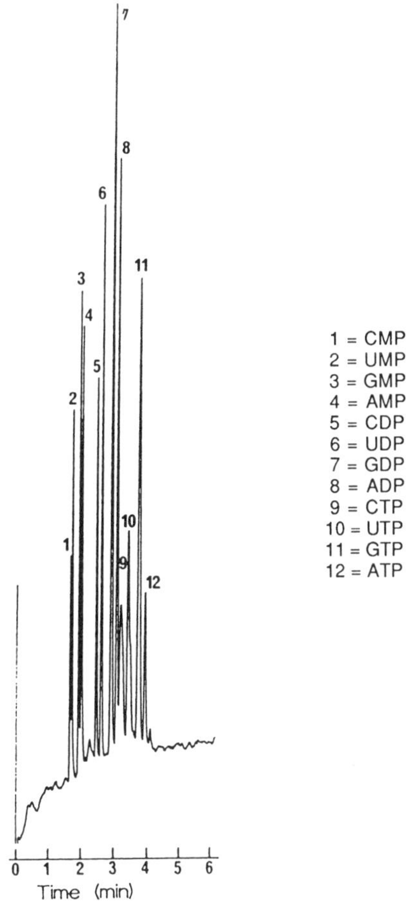

Figure 5.10
MEKC separation of common nucleotides with a cationic surfactant. Column : 50 μm x 26/45 cm. Buffer : 100 mM DTAB in 10 mM TRIS/phosphate buffer, pH 7.05. Applied voltage : -18 kV. Detection : UV, 254 nm. *(Reprinted with permission [43], © MicroSeparations Inc.))*

into the other. Until now, case 10 had little analytical importance, but it has been observed [55,56].

Intermediate forms however can be considered in different ways. *Terabe* classifies case 7 as a form of ion-exchange electrokinetic chromatography [3,57,58], stressing the partitioning between the buffer and a carrier which is not necessarily immobilized, nor forms a distinct phase. According to our classification it resembles more electrophoresis as it can be considered to be related to case 3.

As cyclodextrins form homogeneous solutions [1], there is more hesitance to regard the hydrophobic cavity as a distinct microphase. Use of uncharged cyclodextrins (case 8) has generally been classified as a case of complexation electrophoresis, even after immobilization in a gel [59]. But *Terabe* considers case 9 as a form of electrokinetic inclusion chromatography [3,60], although it lacks the prerequisite of phase separation and notwithstanding its resemblance to case 4. On the other hand the formal description of migration behaviour is very similar as the electrophoretic mobility of the micelle and of the charged cyclodextrin have a comparable role [60].

Table 5.4 Methodologies to obtain electrokinetic separation.

	Analyte	Secondary factor (K_{eq})	$\Delta\mu_{intr}$	K_{eq} charge	K_{eq} mass	Phase separation
1	Ionic[a)]	-	+++			
2	Ionic[b)]	Buffer pH	++	+		
3	Ionic	Charged ligand	+	+	+	
4	Non-ionic	Charged ligand		+++		
5	Ionic	Neutral ligand	+		++	
6	Non-ionic	Neutral ligand		(no separation)		
7	Ionic	Polyion	+	+	+	
8	Ionic	Neutral cyclodextrin	+		++	
9	Non-ionic	Charged cyclodextrin		+++		
10	Ionic	Neutral surfactant	+			++
11	Non-ionic	Neutral surfactant		(no separation)		
12	Ionic	Charged surfactant	+			++
13	Non-ionic	Charged surfactant				+++
14	Ionic	Stationary phase[c)]	+			++
15	Non-ionic	Stationary phase[c)]				+++

a) Strong electrolyte
b) Weak electrolyte
c) Electrochromatography

[1] *Crown ethers, almost exclusively used in isotachophoresis [51], are not included in* ***Table 5.4****, but can be considered similar to cyclodextrins.*

References

[1] S.Terabe, K.Otsuka, K.Ichikawa, A.Tsuchiya, T.Ando, Anal.Chem., **56** (1984) 111.
Electrokinetic separations with micellar solutions and open-tubular capillaries.

[2] S.Terabe, K.Otsuka, T.Ando, Anal.Chem., **57** (1985) 834.
Electrokinetic chromatography with micellar solution and open-tubular capillary.

[3] S.Terabe, Trends Anal.Chem., **8** (1989) 129.
Electrokinetic chromatography : an interface between electrophoresis and chromatography.

[4] T.Nakagawa, Newl., Div.Colloid Surf.Chem., Chem.Soc.Jpn.,**6**(3) (1981) 1.
(Cited in [1-3])

[5] D.E.Burton, M.J.Sepaniak, M.P.Maskarinec, J.Chromatogr.Sci., **24** (1986) 347.
Analysis of B6 vitamers by micellar electrokinetic capillary chromatography with laser-excited fluorescence detection.

[6] K.Otsuka, S.Terabe, J.Microcol.Sep., **1** (1989) 150.
Effect of pH on electrokinetic velocities in micellar electrokinetic chromatography.

[7] M.M.Bushey, J.W.Jorgenson, J.Microcol.Sep., **1** (1989) 125.
Effects of methanol-modified mobile phase on the separation of isotopically substituted compounds by micellar electrokinetic capillary chromatography.

[8] M.M.Bushey, J.W.Jorgenson, Anal.Chem., **61** (1989) 491.
Separation of dansylated methylamine and dansylated methyl-d_3-amine by micellar electrokinetic capillary chromatography with methanol-modified mobile phase.

[9] K.Saitoh, C.Kiyohara, N.Suzuki, J.High Resolut.Chromatogr., **14** (1991) 245.
Mobilities of metal-β-diketonate complexes in micellar electrokinetic chromatography.

[10] K.H.Row, W.H.Griest, M.P.Maskarinec, J.Chromatogr., **409** (1987) 193.
Separation of modified nucleic acid constituents by micellar electrokinetic capillary chromatography.

[11] H.W.Hoyer, K.J.Mysels, D.Stigter, J.Phys.Chem., **58** (1954) 385.
Tracer electrophoresis. I. Free liquid method.

[12] D.Stigter, K.J.Mysels, J.Phys.Chem., **59** (1955) 45.
Tracer electrophoresis. II. The mobility of the micelle of sodium lauryl sulfate and its interpretation in terms of zeta potential and charge.

[13] K.Otsuka, S.Terabe, J.Chromatogr., **515** (1990) 221.
Enantiomeric resolution by micellar electrokinetic chromatography with chiral surfactants.

[14] S.Terabe, O.Shibata, T.Isemura, J.High Resolut.Chromatogr., **14** (1991) 52.
Band-broadening evaluation by back-and-forth capillary electrophoresis.

[15] K.Otsuka, S.Terabe, T.Ando, J.Chromatogr., **348** (1985) 39.
Electrokinetic chromatography with micellar solutions. Retention behaviour and separation of chlorinated phenols.

[16] K.Otsuka, S.Terabe, T.Ando, J.Chromatogr., **332** (1985) 219.
Electrokinetic chromatography with micellar solutions. Separation of phenylthiohydantoin-amino acids.

[17] H.Nishi, N.Tsumagari, S.Terabe, Anal.Chem., **61** (1989) 2434.
Effect of tetraalkylammonium salts on micellar electrokinetic chromatography of ionic substances.

[18] T.Saitoh, H.Hoshino, T.Yotsuyanagi, J.Chromatogr., **469** (1989) 175.
Separation of 4-(2-pyridylazo)-resorcinolato metal chelates by micellar electrokinetic capillary chromatography.

[19] A.T.Balchunas, M.J.Sepaniak, Anal.Chem., **60** (1988) 617.
Gradient elution for micellar electrokinetic capillary chromatography.

[20] A.T.Balchunas, M.J.Sepaniak, Anal.Chem., **59** (1987) 1466.
Extension of elution range in micellar electrokinetic capillary chromatography.

[21] A.C.Powell, M.J.Sepaniak, J.Microcol.Sep., **2** (1990) 278.
Development of a model for predicting retention times in solvent-gradient micellar electrokinetic capillary chromatography.

[22] R.O.Cole, M.J.Sepaniak, W.L.Hinze, J.High Resolut.Chromatogr., **13** (1990) 579.
Optimization of binaphthyl enantiomer separations by capillary zone electrophoresis using mobile phases containing bile salts and organic solvent.

[23] H.Nishi, T.Fukuyama, M.Matsuo, S.Terabe, J.Chromatogr., **513** (1990) 279.
Separation and determination of lipophilic corticosteroids and benzothiazepin analogues by micellar electrokinetic chromatography using bile salts.

[24] J.Vindevogel, P.Sandra, J.High Resolut.Chromatogr., **14** (1991) 795.

Micellar electrokinetic chromatography of ionic compounds : The analysis of hop bitter acids.
[25] J.Vindevogel, P.Sandra, L.C.Verhagen, J.High Resolut.Chromatogr., **13** (1990) 295.
Separation of hop bitter acids by capillary zone electrophoresis and micellar electrokinetic chromatography with UV-diode array detection.
[26] S.Terabe, M.Shibata, Y.Miyashita, J.Chromatogr., **480** (1989) 403.
Chiral separation by electrokinetic chromatography with bile salt micelles.
[27] D.E.Burton, M.J.Sepaniak, M.P.Maskarinec, J.Chromatogr.Sci., **25** (1987) 514.
Evaluation of the use of various surfactants in micellar electrokinetic capillary chromatography.
[28] H.T.Rasmussen, L.K.Goebel, H.M.McNair, J.Chromatogr., **517** (1990) 549.
Micellar electrokinetic chromatography employing sodium alkyl sulfates and Brij 35.
[29] K.Otsuka, S.Terabe, T.Ando, Nippon Kagaku Kaishi, **7** (1986) 950 (Chem.Abstr., **105** (1986) 126484r).
(Separation of aromatic sulfides by electrokinetic chromatography with micellar solutions.)
[30] H.Nishi, N.Tsumagari, T.Kakimoto, S.Terabe, J.Chromatogr., **465** (1989) 331.
Separation of water-soluble vitamins by micellar electrokinetic chromatography.
[31] H.Nishi, N.Tsumagari, T.Kakimoto, S.Terabe, J.Chromatogr., **477** (1989) 259.
Separation of β-lactam antibiotics by micellar electrokinetic chromatography.
[32] A.Dobashi, T.Ono, S.Hara, J.Yamaguchi, J.Chromatogr., **480** (1989) 413.
Enantioselective hydrophobic entanglement of enantiomeric solutes with chiral functionalized micelles by electrokinetic chromatography.
[33] A.Dobashi, T.Ono, S.Hara, J.Yamaguchi, Anal.Chem., **61** (1989) 1984.
Optical resolution of enantiomers with chiral mixed micelles by electrokinetic chromatography.
[34] K.Otsuka, S.Terabe, Electrophoresis (Weinheim), **11** (1990) 982.
Effects of methanol and urea on optical resolution of phenylthiohydantoin-DL-amino acids by micellar electrokinetic chromatography with sodium N-dodecanoyl-L-valinate.
[35] E.A.G.Aniansson, S.N.Wall, M.Almgren, H.Hoffmann, I.Kielmann, W.Ulbricht, R.Zana, J.Lang, C.Tondre, J.Phys.Chem., **80** (1976) 905.
Theory of the kinetics of micelle equilibria and quantitative interpretation of chemical relaxation studies of micellar solutions of ionic surfactants.
[36] A.Roda, A.F.Hofmann, K.J.Mysels, J.Biol.Chem., **258** (1983) 6362.
The influence of bile salt structure on self-association in aqueous solution.
[37] B.Lindman, H.Wennerström, Topics Curr.Chem., **87** (1980) 1.
Micelles. Amphiphile aggregation in aqueous solution.
[38] E.J.Fendler, J.H.Fendler in V.Gold (ed.), Advances in Physical Organic Chemistry, Vol.8, Academic Press, New York, 1970.
Micellar catalysis in inorganic reactions : kinetic and mechanistic implications.
[39] H.Nishi, T.Fukuyama, M.Matsuo, S.Terabe, J.Chromatogr., **498** (1990) 313.
Separation and determination of the ingredients of a cold medicine by micellar electrokinetic chromatography with bile salts.
[40] H.Nishi, T.Fukuyama, M.Matsuo, S.Terabe, J.Microcol.Sep., **1** (1989) 234.
Chiral separation of optical isomeric drugs using micellar electrokinetic chromatography and bile salts.
[41] H.Nishi, T.Fukuyama, M.Matsuo, S.Terabe, J.Chromatogr., **515** (1990) 233.
Chiral separation of diltiazem, trimetoquinol and related compounds by micellar electrokinetic chromatography with bile salts.
[42] T.Kaneta, S.Tanaka, H.Yoshida, J.Chromatogr., **538** (1991) 385.
Improvement of resolution in the capillary electrophoretic separation of catecholamines by complex formation with boric acid and control of electroosmosis with a cationic surfactant.
[43] J.Liu, F.Banks, M.Novotny, J.Microcol.Sep., **1** (1989) 136.
High-speed micellar electrokinetic capillary chromatography of the common phosphorylated nucleosides.
[44] T.Tsuda, J.Liq.Chromatogr., **12** (1989) 2501.
Electroosmotic flow and electric current in capillary electrophoresis.
[45] J.Liu, O.Shirota, M.Novotny, Anal.Chem., **63** (1991) 413.
Capillary electrophoresis of amino sugars with laser-induced fluorescence detection.
[46] K.D.Altria, C.F.Simpson, Chromatographia, **24** (1987) 527.
High voltage capillary zone electrophoresis : Operating parameters effects on electroendosmotic flows and electrophoretic mobilities.
[47] X.Huang, J.A.Luckey, M.J.Gordon, R.N.Zare, Anal.Chem., **61** (1989) 766.
Quantitative determination of low molecular weighth carboxylic acids by capillary zone electrophoresis/conductivity detection.

[48] X.Huang, M.J.Gordon, R.N.Zare, J.Chromatogr., **480** (1989) 285.
Effect of electrolyte and sample concentration on the relationship between sensitivity and resolution in capillary zone electrophoresis using conductivity detection.
[49] J.Liu, K.A.Cobb, M.Novotny, J.Chromatogr., **519** (1990) 189.
Capillary electrophoretic separations of peptides using micelle-forming compounds and cyclodextrins as additives.
[50] J.Tehrani, R.Macomber, L.Day, J.High Resolut.Chromatogr., **14** (1991) 10.
Capillary electrophoresis : an integrated system with a unique split-flow sample introduction mechanism.
[51] J.Snopek, I.Jelínek, E.Smolková-Keulemansová, J.Chromatogr., **452** (1988) 571.
Micellar, inclusion and metal-complex enantioselective pseudophases in high-performance electromigration methods.
[52] S.Honda, S.Iwase, A.Makino, S.Fujiwara, Anal.Biochem., **176** (1989) 72.
Simultaneous determination of reducing monosaccharides by capillary zone electrophoresis as the borate complexes of N-2-pyridylglycamines.
[53] Y.Walbroehl, J.W.Jorgenson, Anal.Chem., **58** (1986) 479.
Capillary zone electrophoresis of neutral organic molecules by solvophobic association with tetraalkylammonium ion.
[54] J.C.Giddings, Anal.Chem., **53** (1981) 945A.
Future pathways for analytical separations.
[55] S.A.Swedberg, J.Chromatogr., **503** (1990) 449.
Use of non-ionic and zwitterionic surfactants to enhance selectivity in high-performance capillary electrophoresis. An apparent micellar electrokinetic capillary chromatographic mechanism.
[56] S.Hjertén, L.Valtcheva, K.Elenbring, D.Eaker, J.Liq.Chromatogr, **12** (1989) 2471.
High-performance electrophoresis of acidic and basic low-molecular-weigth compounds and of proteins in the presence of polymers and neutral surfactants.
[57] S.Terabe, T.Isemura, Anal.Chem., **62** (1990) 650.
Ion-exchange electrokinetic chromatography with polymer ions for the separation of isomeric ions having identical electrophoretic mobilities.
[58] S.Terabe, T.Isemura, J.Chromatogr., **515** (1990) 667.
Effect of polymer ion concentrations on migration velocities in ion-exchange electrokinetic chromatography.
[59] A.Guttman, A.Paulus, A.S.Cohen, N.Grinberg, B.L.Karger, J.Chromatogr., **448** (1988) 41.
Use of complexing agents for selective separation in high-performance capillary electrophoresis. Chiral resolution via cyclodextrins incorporated within polyacrylamide gel columns.
[60] S.Terabe, H.Ozaki, K.Otsuka, T.Ando, J.Chromatogr., **332** (1985) 211.
Electrokinetic chromatography with 2-O-carboxymethyl-β-cyclodextrin as a moving 'stationary' phase.

6 Thermal Effects in Capillary Electrophoretic Techniques

6.1 High-Voltage Sources

In classical electrophoresis, the applied voltage is generally limited to less than 10 kV. In capillary electrophoresis it can be as high as 30 kV, but developed power is reduced as currents are much smaller, typically below 100 μA. Nevertheless, components that are subjected to high voltages are preferably enclosed in a Plexiglass housing, equipped with a safety switch that turns off the voltage on opening. Applied fields are limited to a few hundreds of volts/cm, but, with a special column arrangement, values as high as 5000 V/cm have been used [1].

Under favorable conditions (good temperature regulation, stabilized column, sufficient buffer capacity ...) currents can be very stable. The profile reproduced as an example in **Figure 6.1** shows less than 0.2 % relative standard deviation around the average value of 25.8 μA (not counting the initial ramp). Drift, over a 5 minute period, is less than 0.03 μA/min. Especially with automated instrumentation, the possibility of continuously monitoring the current is a useful option, allowing checks for anomalies like drift or sudden changes. As will be later discussed, if current instability is caused by temperature variation, application of constant current instead of constant voltage may be preferred.

Usually, the high voltage is applied on the injection side of the column, with the detector side grounded. This arrangement reduces the risk of damage to the detector, and also produces less noise in detectors that are sensitive to voltage-induced sparks. Electrochemical detectors, but also the frequently used optical detectors do show this increased noise and can require additional grounding of parts that are in contact with the capillary to eliminate it. Therefore, reversal of polarity is generally performed by changing the sign of the applied voltage (in some cases this requires another voltage source), while keeping the ground at the detector side (**Figure 6.2**). Occasionally, high-voltage is applied at the detector side (**Figure 6.2c**), for instance when special injection systems for single-cell analysis [2] or continuous buffer feed for gradient elution [3,4] are used. This can require electric insulation of the detector with Plexiglass slabs [3,4].

Stabilized high-voltage sources, with low residual ripple ($\leq$ 0.1%) and preferably with options for polarity reversal and constant current mode, will suffice for most CZE and MEKC applications. Only occasionally, more sophisticated means for applying voltage are necessary. One example is the agarose gel electrophoresis of large DNA fragments. In classical electrophoresis special non-uniform electric field methods have been developed [5,6]. They are based on field-dependent mobility differences (reptation), using periodic changes in the direction of the applied field. Given the one-dimensional nature of capillary techniques, such changes are limited to full field reversals. This technique has been used for the analysis of DNA digests [7].

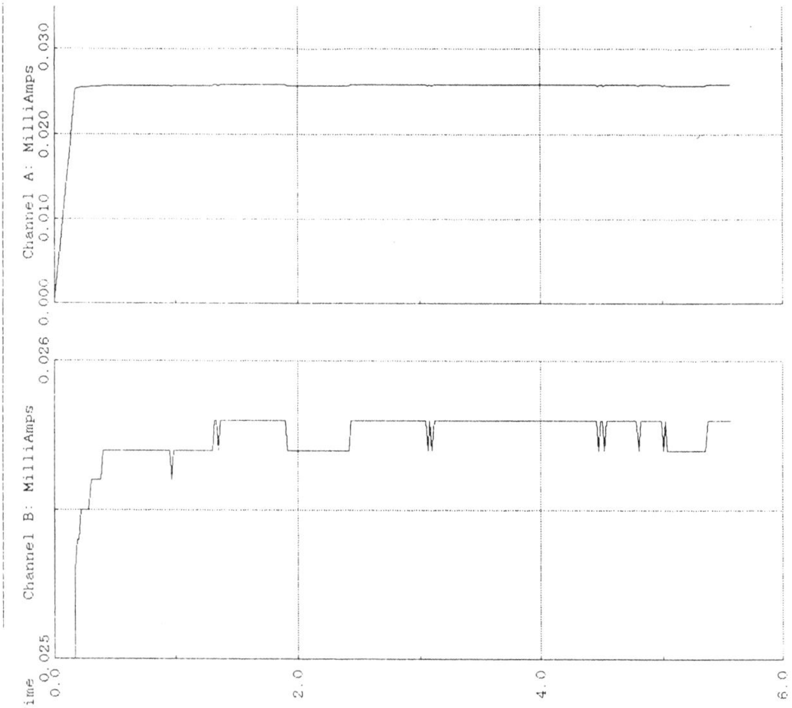

Figure 6.1
Current profile as a function of time. Capillary : 75 µm x 43/50 cm. Buffer : 25 mM Tris/HCl, pH = 8.15. Applied voltage 20 kV (at 25°C, liquid cooled capillary).

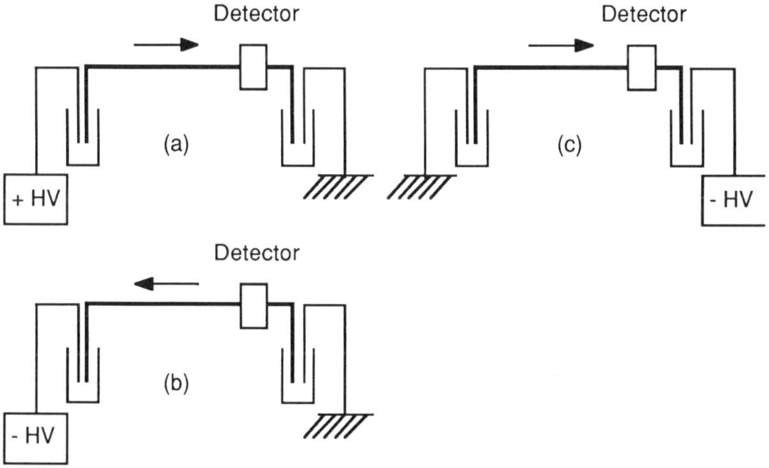

Figure 6.2
Applying high-voltage (arrows indicate electroosmotic flow under usual conditions, fused silica capillary with neutral or alkaline buffer).
(a) to (b) : Changing polarity and maintaining 'hot' side at injector.
(a) to (c) : Maintaining polarity but moving 'hot' side to detector.

6.2 Thermal Equilibrium in Capillaries

During the electrophoretic process, the electrokinetic energy is released as heat : the Joule effect. Part of the thermal energy is transmitted to the environment, but part of it is consumed in a temperature rise of the medium. Ion mobilities, and thus current, are a function of the viscosity. As the temperature of the solution is raised, viscosity is decreased, allowing a further increase in current, and also in released heat. (**Figure 6.3**). If heat removal to the environment is insufficient, a thermal 'catastrophe' [8] can occur. The temperature rises so high that the solution starts to boil and the subsequent formation of bubbles will inhibit further passage of current.

The aim of miniaturization, as applied in capillary zone electrophoresis, is to increase the surface-to-volume ratio, so that heat transfer to the environment is more effective. For practical reasons, capillaries with a circular cross-section are used, although it has been demonstrated that, for equal volumes, flat, rectangular capillaries perform better [9,10]. This is in fact the geometry of the classical slab gels. Such capillaries are commercially available [11], but as their short side has the same dimensions of available cylindrical capillaries, the volume is effectively increased. Due to their fragility, and in spite of the opportunities for optical detection, they are rarely used.

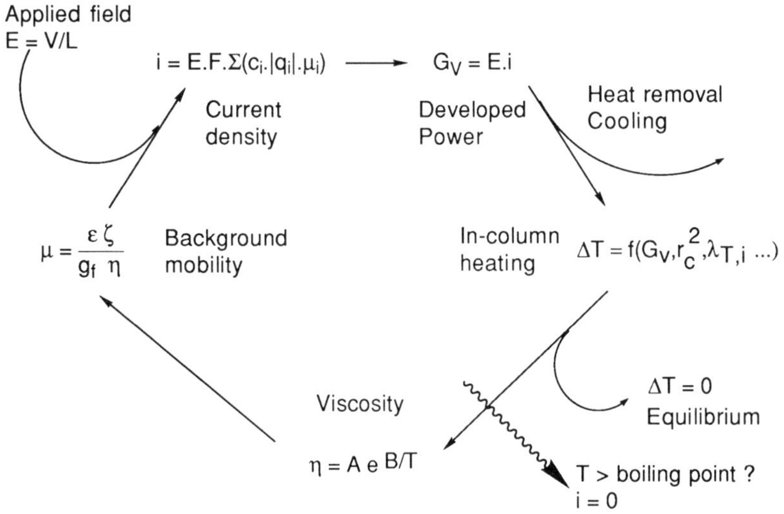

Figure 6.3
Self-enforcing influence of current on developed power during electrophoresis.

6.3 Cooling of Capillaries

The power developed in a capillary, G_V, is determined by the applied field E, the equivalent conductivity Λ of the buffer and the buffer concentration c.

$$G_V = \frac{V.I}{\pi r_c^2 L} = E^2 . \Lambda . c \qquad (6.1)$$

G_V can be several hundreds of watt/cm^3. Due to heat exchange with the surroundings, a temperature profile is established accross the capillary cross-section (**Figure 6.4**). Depending upon the circumstances, the temperature rise in the capillary can be several tens of degrees. Heat conductivity is lowest at the outside of the capillary, especially in the case of air-cooling, hence the importance of convection for cooling. Due to heating, convection will spontaneously develop in the surroundings. However, this natural convection will not always be sufficient to avoid overheating and eventual current breakdown. At particular places, e.g. where the capillary is mounted in the detector, the convection can be seriously hindered.

When high field strengths (> 200 V/cm) are used, active cooling is essential. It has been shown that a strong air stream (> 10 m/s) can diminish the temperature rise in the capillary by a factor 5 [12]. Alternatively, liquid cooling can be applied, which has been reported [13-15] to be more efficient. But even in cooled systems, a temperature rise can be deduced from the non-linearity between current and applied voltage (**Figure 6.5**). The same effect, but on a smaller scale, can be observed when a constant-temperature liquid coolant is used. Therefore, these systems should be characterized as regulating the temperature rather than keeping the temperature constant.

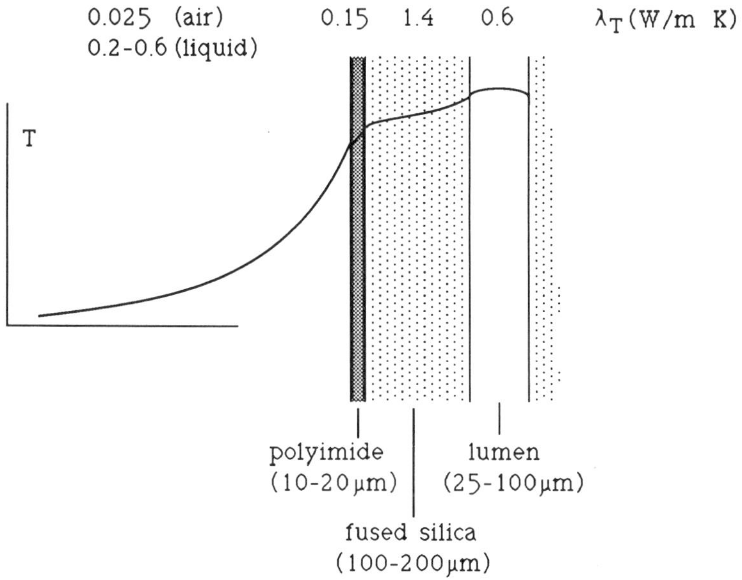

Figure 6.4
Heat conductivity and temperature in- and outside of the capillary.

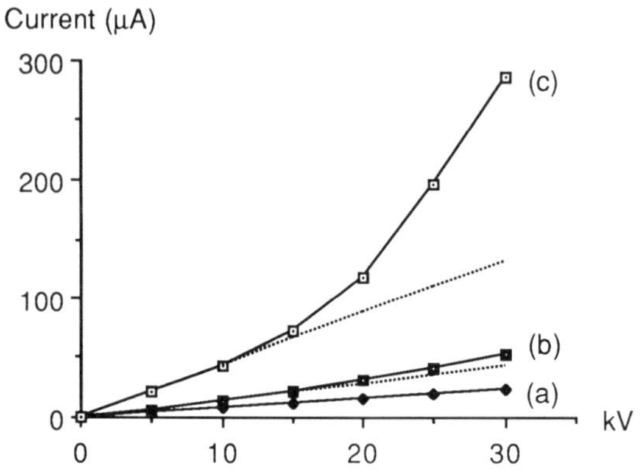

Figure 6.5
Current as a function of applied voltage. Capillary : 75 μm x 60 cm, air-cooled.
(a) 6.25 mM borax, (b) 12.5 mM borax, (c) 40 mM sodium dodecyl sulfate (SDS) in 30 mM Na_2HPO_4/NaH_2PO_4.

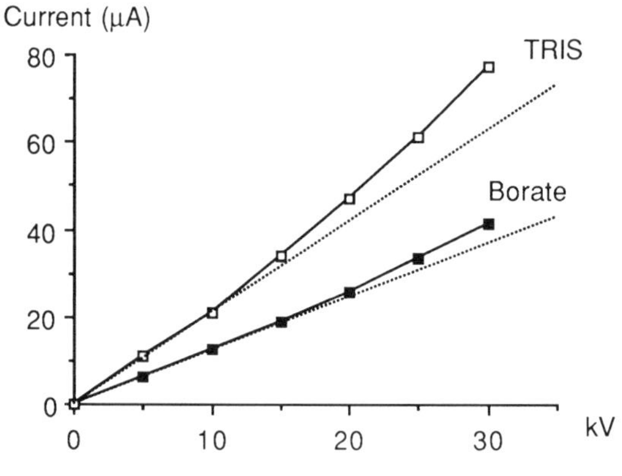

Figure 6.6
Current as a function of applied voltage. Capillary : 75 μm x 57 cm, liquid-cooled.
Buffer : 30 mM SDS in 25 mM borate (pH 8.55) or 25 mM TRIS/HCl (pH 8.4).
(Reprinted with permission [16], © Hüthig Verlag)

6.4 Effect on Mobility and Migration Times

The temperature dependence of the mobility, and consequently of the migration times, is mainly determined by the viscosity :

$$\mu = \frac{\varepsilon \zeta}{g_f \eta} \tag{6.2}$$

and

$$t_M = \frac{g_f \eta . L^2}{\varepsilon \zeta . V} \tag{6.3}$$

Therefore, mobilities will increase by approximately 2%/°. This is not necessarily a matter of concern : in some cases it could reduce resolution, but in others it will even allow faster analysis. More important is that reproducibility of migration times will be determined by the way in which viscosity, and thus temperature, can be held constant during analysis. Again, active cooling is important. Constant, reproducible cooling conditions are essential which can require maintaining constant coolant temperature (e.g. with Peltier elements [13]).

If one does not want to determine certain quantities as a function of the applied voltage, and reproducibility of migration times is considered more important, it can be advantageous to perform the analysis at constant current [17,18]. Mobilities of the buffer ions (current) and analyte ions (migration times) are influenced by the viscosity in the same way. By externally imposing a constant current, an option which is available on most high-voltage sources, the viscosity is actually held constant. For the same reason, reciprocal migration times (1/t) and observed current are better linearly correlated than (1/t) and applied voltage (**Figure 6.7**).

Time-dependent variations of temperature and viscosity also explain why peak areas are less reproducible than migration times [15]. Migration times reflect average conditions over a relatively long time. Peak areas, on the other hand, are determined by the actual conditions during relatively short periods : the mobility during injection (by electromigration) or the velocity with which the analyte passes through the detector.

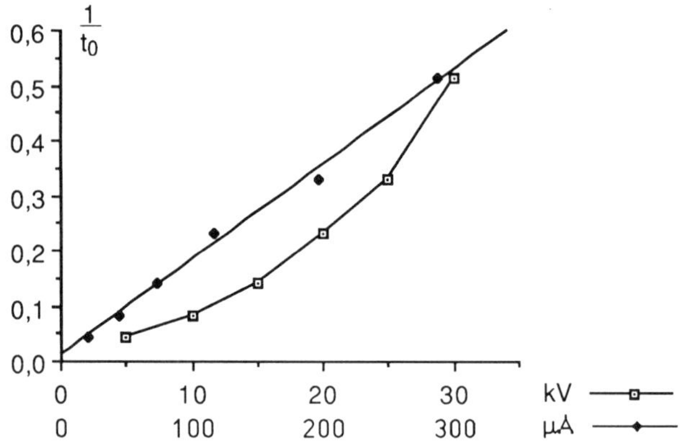

Figure 6.7
Relation between electroosmotic migration time and observed current or applied voltage. Conditions : see **Figure 6.5**, buffer (c).

6.6 Effect on Efficiency

The temperature increase inside the column causes the formation of a temperature gradient between center and wall. At the center, temperature is higher, and by consequence the viscosity is lower. Therefore ions in the center will move faster than near the wall. This causes additional band broadening.

$$\sigma^2 = 2\,D_m.t + \sigma^2_{\Delta T} \qquad (6.5)$$

The exact shape of the gradient and the associated flow profile, and the influence on efficiency N and plate height h have been evaluated [8-10,12,19-25]. Under usual conditions the gradient is well approximated by a parabola [22] and the influence on efficiency is marginal when the temperature difference between center and inner wall are limited to 1 or 1.5 ° [13,21,23].

Knox has formulated a criterion based on the evaluation of the thermal contribution to plate height, assuming that it should be less than 10% of the contribution from axial diffusion.

$$d_c\,(\mu m)\;.\;E\,(kV/m)\;.\;\sqrt[3]{c}\,(mol/l) < 1500 \qquad (6.6)$$

However, the limiting value thus obtained depends on the choice of 'typical' values for diffusion coefficient, mobility and equivalent conductivity of the buffer. Although no prior experimentation is needed in this way, the universal applicability of this criterion is limited. More generally, and if current values are available, a limit can be imposed upon the temperature gradient [13].

$$\Delta T_c = G_V r_c^2 / 4 \lambda_T < 1°C \qquad (6.7)$$

with $\lambda_T = 0.6$ W/m K and $G_V = E.I / \pi r_c^2$

or $E.I < 7.5$ W/m (6.8)

Column diameter and buffer concentration are decisive for the thermal contribution to plate height [23]. With relatively high buffer concentrations (100 mM and $\Lambda = 0.015$ m^2/mol.Ω), the column diameter is preferably limited to 50 µm, as larger diameters will cause loss of efficiency. When the buffer concentration is lowered ten-fold, column diameters up to 125 µm will give satisfactory efficiency, at least for large molecules (low diffusion coefficient). In that case, however, sample concentration must be reduced accordingly to avoid peak broadening by electrokinetic dispersion [26], which will in turn affect detectability. In the case of UV absorbance detection, the optical path length is determined by the column diameter. When high detectability is important, the choice of column diameter will be a compromise between path length, allowable sample concentration, effect of efficiency on resolution and effect of efficiency on peak height.

6.7 Other Thermal Effects

After injection, the sample is situated in a small zone at one of the column extremities. When voltage is applied, the in-column heating will cause the buffer to expand, which in turn pushes the sample zone partly out of the column. Although the sample zone will migrate back into the column, dilution may have occurred, resulting in loss of efficiency. This can be avoided by applying voltage in a step-wise manner [27,28]. In this way, the sample is allowed to migrate in the column before the maximum temperature increase is attained.

Chemical equilibria are influenced by temperature. The actual in-column pH of a buffer can differ from the one measured at room temperature. Especially buffers of the BH^+/B-type, like TRIS/HCl have a temperature-sensitive pK_a. Between 20 and

40°C, the pH can be lowered by half a pH-unit (**Figure 6.8**). As the mobility of weak electrolytes is determined by the degree of dissociation, sample components can also show temperature-sensitive mobilities.

Additives like cyclodextrins, surfactants or complexing agents are used to influence mobilities in a selective way. In all these cases the equilibria involved can be influenced by temperature differences, causing changes in relative mobilities. In MEKC, this can lead to selectivity changes, especially when ionic equilibria are involved. In the separation of hop acid salts, shown in **Figure 6.9**, the applied voltage can be safely increased from 10 to 15 kV without sacrificing much of the resolution of adlupulone (the small marked peak) and other components that elute close to it. In a TRIS buffer, however, selectivity changes lead to a complete loss of resolution between adlupulone and cohumulone. The difference between TRIS and borate buffer results both from higher in-column temperature for TRIS (**Figure 6.6**) and from the stronger pH dependence (**Figure 6.8**). Variance of equilibria along the column diameter also affects efficiency which shows a pronounced fall with applied voltage in a TRIS buffer but not in the borate buffer [16].

Finally, sample components themselves can be temperature sensitive. Protein denaturation caused by in-column heating has been reported [13,27].

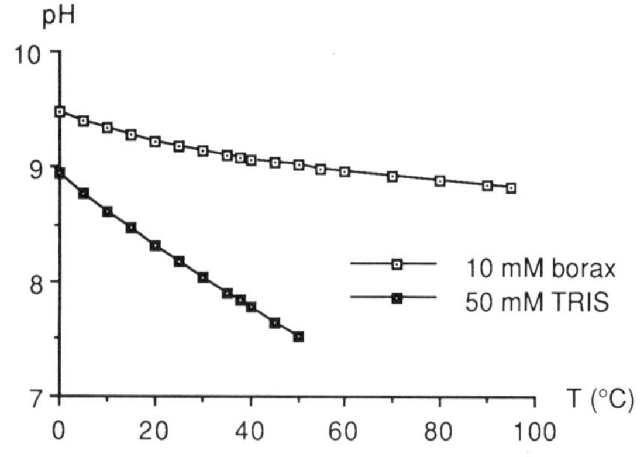

Figure 6.8
Effect of temperature on buffer pH (based on data from [29])

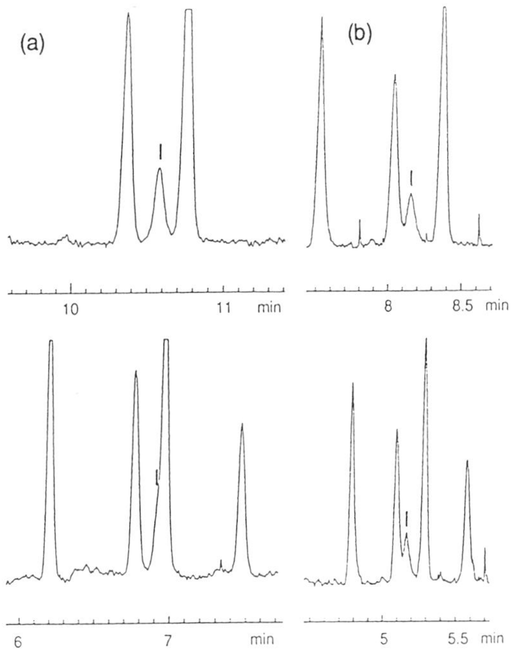

Figure 6.9
Loss of resolution caused by in-column heating in the MEKC analysis of ionic compounds. Capillary : 75 µm x 43/50 cm (liquid-cooled). Buffer : 30 mM SDS in 25 mM TRIS/HCl, pH 8.40 (a) or borate, pH 8.55 (b). Applied voltage : 10 kV (top) or 15 kV (bottom). *(Reprinted with permission [16], © Hüthig Verlag)*

References

[1] C.A.Monnig, J.W.Jorgenson, Anal.Chem., **63** (1991) 802.
On-column sample gating for high-speed capillary zone electrophoresis.
[2] R.T.Kennedy, M.D.Oates, B.R.Cooper, B.Nickerson, J.W.Jorgenson, Science, **246** (1989) 57.
Microcolumn separations and the analysis of single cells.
[3] A.C.Powell, M.J.Sepaniak, J.Microcol.Sep., **2** (1990) 278.
Development of a model for predicting retention times in solvent-gradient micellar electrokinetic capillary chromatography.
[4] M.J.Sepaniak, D.F.Swaile, A.C.Powell, J.Chromatogr., **480** (1989) 185.
Instrumental developments in micellar electrokinetic capillary chromatography.
[5] H.J.S.Dawkins, J.Chromatogr., **492** (1989) 615.
Large DNA separation using field alternation agar gel electrophoresis.
[6] K.Gardiner, Anal.Chem., **63** (1991) 658.
Pulsed field gel electrophoresis.
[7] D.N.Heiger, A.S.Cohen, B.L.Karger, J.Chromatogr., **516** (1990) 33.
Separation of DNA restriction fragments by high performance capillary electrophoresis with low

and zero cross-linked polyacrylamide using continuous and pulsed electric fields.
[8] W.A.Gobie, C.F.Ivory, J.Chromatogr., **516** (1990) 191.
Thermal model of capillary electrophoresis and a method for counteracting thermal band broadening.
[9] J.O.N.Hinckley, J.Chromatogr., **109** (1975) 209.
Electrophoretic thermal theory. I. Temperature gradients and their effects.
[10] M.Jansson, Å.Emmer, J.Roeraade, J.High Resolut.Chromatogr.Chromatogr. Commun., **12** (1989) 797.
Some design considerations in miniaturized electrokinetic separation systems.
[11] T.Tsuda, J.V.Sweedler, R.N.Zare, Anal.Chem., **62** (1990) 2149.
Rectangular capillaries for capillary zone electrophoresis.
[12] J.H.Knox, Chromatographia, **26** (1988) 329.
Thermal effects and band spreading in capillary electro-separation.
[13] R.J.Nelson, A.Paulus, A.Guttman, A.S.Cohen, B.L.Karger, J.Chromatogr., **480** (1989) 111.
Use of Peltier thermoelectric devices to control column temperature in high-performance capillary electrophoresis.
[14] V.P.Burolla, S.L.Pentoney, R.Zare, Am.Biotechn.Lab., **7**(10) (1989) 20.
High-performance capillary electrophoresis.
[15] Y.Kurosu, K.Hibi, T.Sasaki, M.Saito, J.High Resolut.Chromatogr., **14** (1991) 200.
Influence of temperature control in capillary electrophoresis.
[16] J.Vindevogel, P.Sandra, J.High Resolut.Chromatogr., **14** (1991) 795.
Micellar electrokinetic chromatography of ionic compounds : The analysis of hop bitter acids.
[17] S.Terabe, K.Otsuka, T.Ando, Anal.Chem., **57** (1985) 834.
Electrokinetic chromatography with micellar solution and open-tubular capillary.
[18] T.Tsuda, J.Liq.Chromatogr., **12** (1989) 2501.
Electroosmotic flow and electric current in capillary electrophoresis.
[19] S.Hjertén, Chromatogr.Rev., **9** (1967) 122.
Free zone electrophoresis.
[20] M.Coxon, M.J.Binder, J.Chromatogr., **101** (1974) 1.
Radial temperature distribution in isotachophoresis columns of circular cross-section.
[21] J.H.Knox, I.W.Grant, Chromatographia, **24** (1987) 135.
Miniaturisation in pressure and electroendosmotically driven liquid chromatography : Some theoretical considerations.
[22] A.E.Jones, E.Grushka, J.Chromatogr., **466** (1989) 219.
Nature of temperature gradients in capillary zone electrophoresis.
[23] E.Grushka, R.M.McCormick, J.J.Kirkland, Anal.Chem., **61** (1989) 241.
Effect of temperature gradients on the efficiency of capillary zone electrophoresis separations.
[24] J.M.Davis, J.Chromatogr., **517** (1990) 521.
Influence of thermal variation of diffusion coefficient on non-equilibrium plate height in capillary zone electrophoresis.
[25] F.Foret, M.Deml, P.Boček, J.Chromatogr., **452** (1988) 601.
Capillary zone electrophoresis. Quantitative study of the effects of some dispersive processes on the separation efficiency.
[26] F.E.P.Mikkers, F.M.Everaerts, Th.P.E.M.Verheggen, J.Chromatogr., **169** (1979) 1.
Concentration distributions in free zone electrophoresis.
[27] R.M.McCormick, Anal.Chem., **60** (1988) 2322.
Capillary zone electrophoretic separation of peptides and proteins using low pH buffers in modified silica capillaries.
[28] M.M.Bushey, J.W.Jorgenson, J.Microcol.Sep., **1** (1989) 125.
Effects of methanol-modified mobile phase on the separation of isotopically substituted compounds by micellar electrokinetic capillary chromatography.
[29] G.D.Fasman (ed), "Handbook of Biochemistry and Molecular Biology, Physical and Chemical Data, Vol.1", CRC Press, Cleveland, OH, USA (1976).

7 Columns

7.1 Column Material and Miniaturization

Borosilicate glass was used in the early days of CZE [1,2], but its high UV cut-off (280 nm, [3]) is an obvious disadvantage when on-column optical detection techniques are used. As much of the early work was done in isotachophoretic apparatus, Teflon capillaries have also been used [4]. Apart from the relatively large diameters (generally > 100 μm), Teflon capillaries also show other disadvantages. They are certainly not inert : they do show electroosmotic flow [5,6], indicative of charge separation between column wall and buffer, and they do adsorb proteins [7]. Possibly, this activity is related to the reported breakdown of C-F bonds under the influence of applied high voltages [8]. Due to poorer thermal conductivity, they also tend to overheat more easily [7]. In this respect, the superiority of silicon columns with high thermal conductivity has been discussed theoretically [9], but the consequences of applying several thousands of volts over a semi-conductor were not taken into account.

Nowadays, fused silica capillaries [10] are used for the majority of applications, generally with internal diameters of 50 or 75 μm. Especially when high mass-sensitivity is desired, smaller diameters are applied, even down to 2 μm [11], requiring dedicated detection methods. The consequences of miniaturization for the volumetric properties of columns, used in capillary electrophoresis can be appreciated from some calculated data summarized in **Table 7.1**.

Table 7.1 Some volumetric column properties as a function of internal diameter

Column diameter	μm	2	5	10	25	50	75	100	200
Volume	μl/m	0.003	0.02	0.08	0.49	1.96	4.42	7.85	31.42
Flow rate a)	nl/min	0.19	1.18	4.71	29.5	118	265	471	1885
Surface/volume	μm^{-1}	2.00	0.80	0.40	0.16	0.08	0.05	0.04	0.02

a) For an electroosmotic velocity of 1 mm/sec

7.2 Column Surfaces

It was recognized very early [1] that with the increase of the surface to volume ratio, necessary for effective cooling, sorption problems can occur. Today, surface and surface adsorption problems are acknowledged to be of major concern for further development of capillary electrokinetic techniques.

Surfaces affect separations in two different ways. First, they determine the electroosmotic flow, and as this can form a considerable component of the analyte linear velocity, the reproducibility of the electroosmotic flow, or the lack of it, will determine reproducibility of the analysis. The second problem is wall adsorption. If it is irreversible then it can lead to loss of sample components. But even if reversible, the chromatographic partitioning with the wall can lead to serious peak broadening and loss of efficiency. Proteins, especially those with a high isoelectric point, are very liable to adsorption[1] and it has been predicted that a column wall-related capacity factor as low as 0.05 could reduce plate numbers 20-fold [12].

The two phenomena are obviously linked as the charge separation, responsible for electroosmosis, also delivers the active sites for adsorption. Adsorbed species can then modify the double layer structure, introducing short term and local or long term variability in the electroosmotic flow. Therefore, elimination of electroosmosis and adsorption often, although not always, go hand in hand. Most of the surface deactivation studies are closely related to the CZE analysis of proteins. In MEKC, however, analytes and analytical conditions differ considerably. The usefulness of coatings developed for the CZE of proteins is not evident in MEKC. With small organic molecules, wall adsorption can actually be promoted by the presence of an organic coating. In MEKC, the use of coatings is more concerned with the control of the electroosmotic flow and its consequences for resolution optimization. This topic will be discussed in Part III. The next paragraph just summarizes some deactivation methods that have been developed and which are applied in CZE.

7.3 Surface Deactivation Methods in CZE

Column wall condition can be controlled through buffer composition or by permanent modification.

7.3.1 Dynamic Deactivation Methods

In fused silica columns, extreme buffer pH values can be applied to avoid

[1] *The interactions of proteins with silica surfaces have been reviewed [12].*

protein adsorption (**Figure 7.1**). Above pH 2 the surface becomes negatively charged. On further raising the pH to a value above the isoelectric point, the protein is also negatively charged and will be repelled by the wall [13]. Alternatively, the pH can be lowered below the isoelectric point, thereby reducing the dissociation of the silanol groups and also the electroosmotic flow [14]. Some limitations must be considered : some proteins may denature under extreme pH conditions, the freedom in pH selection for optimizing mobility differences is restricted and some proteins like lysozyme (pI = 11.0) are too basic for analysis at high pH. High ionic strength buffers have also been found useful for preventing protein adsorption but the high background conductivity can cause problems [12,15]. Diamines have also been used to reduce surface activity [13,16-18]. Recently, the dynamic coating of the wall with an 'omniphobic' fluorosurfactant has been described, allowing the separation of basic proteins at neutral pH [19].

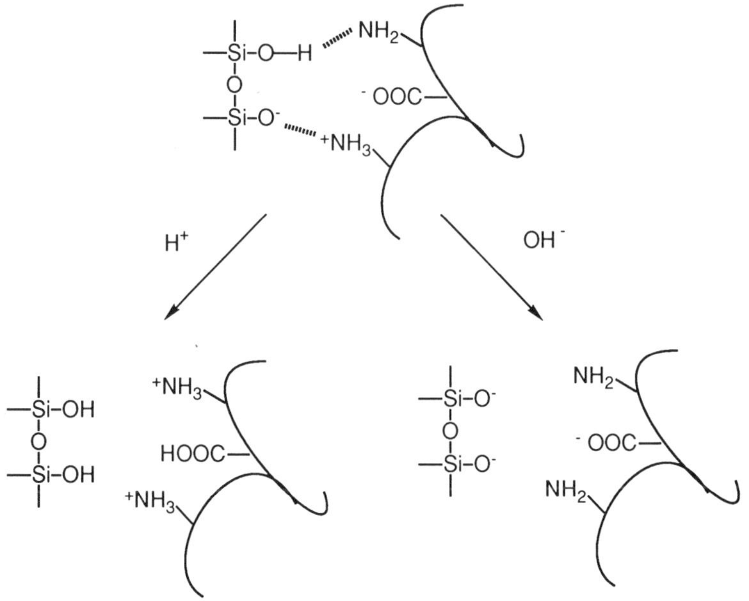

Figure 7.1
Rationale for using extreme pH buffers to avoid protein adsorption.

7.3.2 Static Methods to Avoid Wall Adsorption : Coated Columns

It should be stressed that column coatings are not applied for chromatographic reasons. On the contrary, they are used to avoid adsorption or partition between wall and buffer. Although much of the chemistry, in composition as well as in coating procedure, is borrowed from chromatographic column technology, reference to 'bonded phases' [20,21] is obscuring and should therefore be avoided.

A great variety of coating materials has been proposed (**Table 7.2**). Resistance to hydrolysis is important as the working pH-range in capillary electrokinetic analysis is larger than in LC : values can be as low as 2 but also as high as 11. Operating principles also differ : shielding of the fused silica surface from polar interaction (non-ionic coatings) or charge repulsion between wall and solute (ionic coatings). Both methods assume a particular relationship between coating and analyte. They can have contrary effects to the one intended if the analyte is hydrophobic or has a charge opposite to the coating. Therefore, it is not likely that a single, universal coating will solve all problems.

Table 7.2 Coatings used in the modification of fused silica capillaries.

Non-ionic coatings	
Carbohydrate-related coatings	
Methylcellulose	[1,22]
Maltose	[23]
Acrylic and related coatings	
Polyacrylamide	[24,25]
Polyacrylic acid	[14]
Polyvinylpyrrolidone	[14]
Siloxanes	[1,26,27,28]
Diols, Polyols, Polyethylene Glycol	
Polyols a)	[7,14]
Polyethylene glycol	[22,29]
Others	
Pentafluorobenzoyl coatings	[30]
Ionic Coatings	
Aminopropylsilylation	[20,31]
Sulfonic acid based cation exchangers	[32]
Polyethylene imine	[33]
Physically adsorbed cationic polymer b)	[34]

a) Obtained by hydrolysis and related ring-opening reactions of epoxy-coated fused silica.
b) Of non-disclosed nature.

Carbohydrate coatings are effective in eliminating electroosmosis [1,22] although not very stable. Not only desorption, but also microbial degradation affects coating performance, as demonstrated by *Bruin* et al. [23]. Their maltose-based coating required addition of 0.01% sodium azide to the buffer to prevent breakdown. This coating showed some (reversed) electroosmosis, ascribed to remaining aminopropyl groups, used for fixation to the fused silica.

Linear polyacrylamide coatings have been developed to counteract electroosmosis in relatively wide capillaries (3 mm and 200 µm ID), in which efficiency is lost by hydrodynamic backflow [24]. For small anionic molecules, the coating proved to be superior to methylcellulose coating. However, in more conventional columns (75 µm), it was found that such a coating causes a loss in efficiency for the CZE analysis of organic anions [35]. Interaction of the anions is more likely with the organic coating than with the negatively charged silica surface. This is one example of how a coating can actually be misused.

Hydrophobic silicone coatings reduce but do not eliminate electroosmosis completely [1]. Whether this is due to incomplete coverage of the silica surface has not been definitely demonstrated, but *Sepaniak* [27] reported the necessity of a 'break-in' period, involving a few protein injections, before conalbumin could be analyzed on an octadecylated column, even at a pH above the pI.

Polyethylene glycol (PEG) coatings provide long-lasting reduction in electroosmotic flow [22].

The use of ionic coatings is based on charge repulsion rather than shielding of active sites.They also can provide electroosmotic flow where desirable. Depending upon the type of coating, the variation of the electroosmotic flow with the pH can be modified or reversed. Using sulfonic acid based, strong cation exchangers reduces the pH dependence although small contributions from surface silanol groups are still recognized [32]. Aminopropylsilylation (APS), on the other hand, reverses the charge on the wall and consequently the direction of the electroosmotic flow [22].

References

[1] S.Hjertén, Chromatogr.Rev., **9** (1967) 122.
Free zone electrophoresis.

[2] J.W.Jorgenson, K.D.Lukacs, Anal.Chem., **53** (1981) 1298.
Zone electrophoresis in open-tubular glass capillaries.

[3] J.W.Jorgenson, K.D.Lukacs, Science, **222** (1983) 266.
Capillary zone electrophoresis.

[4] F.E.P.Mikkers, F.M.Everaerts, Th.P.E.M.Verheggen, J.Chromatogr., **169** (1979) 11.
High-performance zone electrophoresis.

[5] J.C.Reijenga, G.V.A.Aben, Th.P.E.M.Verheggen, F.M.Everaerts, J.Chromatogr., **260** (1983) 241.
Effect of electroosmosis on detection in isotachophoresis.

[6] K.D.Lukacs, J.W.Jorgenson, J.High Resolut.Chromatogr.Chromatogr.Commun., **8** (1985) 407.
Capillary zone electrophoresis : Effect of physical parameters on separation efficiency and quantitation.

[7] J.W.Jorgenson, Trends Anal.Chem., **3** (1984) 51.
Zone electrophoresis in open-tubular capillaries.

[8] A.Zhu, Y.Chen, J.Chromatogr., **470** (1989) 251.
High-voltage capillary zone electrophoresis of red blood cells.

[9] M.Jansson, Å.Emmer, J.Roeraade, J.High Resolut.Chromatogr. Chromatogr. Commun., **12** (1989) 797.
Some design considerations in miniaturized electrokinetic separation systems.

[10] R.Dandeneau, E.H.Zerenner, J.High Resolut.Chromatogr. Chromatogr. Commun., **2** (1979) 35
An investigation of glasses for capillary chromatography.

[11] A.G.Ewing, T.M.Olefirowicz, Anal.Chem., **62** (1990) 1872.
Capillary electrophoresis in 2 and 5 μm diameter capillaries : application to cytoplasmatic analysis.

[12] J.S.Green, J.W.Jorgenson, J.Chromatogr., **478** (1989) 63.
Minimizing adsorption of proteins on fused silica in capillary zone electrophoresis by the addition of alkali metal salts to the buffer.

[13] H.H.Lauer, D.McManigill, Anal.Chem., 58 (1986) 166.
Capillary zone electrophoresis of proteins in untreated fused silica tubing.

[14] R.M.McCormick, Anal.Chem., **60** (1988) 2322.
Capillary zone electrophoretic separation of peptides and proteins using low pH buffers in modified silica capillaries.

[15] M.M.Bushey, J.W.Jorgenson, J.Chromatogr., **480** (1989) 301.
Capillary electrophoresis of proteins in buffers containing high concentrations of zwitterionic salts.

[16] F.S.Stover, B.L.Haymore, R.J.McBeath, J.Chromatogr., **470** (1989) 241.
Capillary zone electrophoresis of histidine-containing compounds.

[17] V.Rohlíček, Z.Deyl, J.Chromatogr., **494** (1989) 87.
Simple apparatus for capillary zone electrophoresis and its application to protein analysis.

[18] J.A.Bullock, L.-C.Yuan, J.Microcol.Sep., **3** (1991) 241.
Free solution capillary electrophoresis of basic proteins in uncoated fused silica capillary tubing.

[19] Å.Emmer, M.Jansson, J.Roeraade, J.Chromatogr., **547** (1991) 544.
Improved capillary zone electrophoretic separation of basic proteins, using a fluorosurfactant buffer additive.

[20] C.Wooley, A.Dougherty, K.Duff, Proceedings of the 13th Int.Symp. on Capillary Chromatography, Riva del Garda, Italy (1991) 1189.
Novel proprietary bonded phases for capillary electrophoresis.

[21] Y.-F.Maa, K.J.Hyver, S.A.Swedberg, J.High Resolut.Chromatogr., **14** (1991) 65.
Impact of wall modifications on protein elution in high-performance capillary zone electrophoresis.

[22] B.J.Herren, S.G.Shafer, J.van Alstine, J.M.Harris, R.S.Snyder, J.Colloid Interface Sci., 115 (1987) 46.
Control of electroosmosis in coated quartz capillaries.

[23] G.J.M.Bruin, R.Huisden, J.C.Kraak, H.Poppe, J.Chromatogr., **480** (1989) 339.
Performance of carbohydrate-modified fused-silica capillaries for the separation of proteins by zone electrophoresis.

[24] S.Hjertén, J.Chromatogr., **347** (1985) 191.
High-performance electrophoresis. Elimination of electroendosmosis and solute adsorption.

[25] K.A.Cobb, V.Dolnik, M.Novotny, Anal.Chem., **62** (1990) 2476.

Electrophoretic separations of proteins in capillaries with hydrolytically stable surface structures.
[26] W.G.Kuhr, E.S.Yeung, Anal.Chem., **60** (1988) 2642.
Optimization of sensitivity and separation in capillary zone electrophoresis with indirect fluorescence detection.
[27] M.J.Sepaniak, D.F.Swaile, A.C.Powell, R.O.Cole, J.High Resolut.Chromatogr., **13** (1990) 679.
Capillary electrokinetic separations : influence of mobile phase composition on performance.
[28] J.K.Towns, F.E.Regnier, Anal.Chem., **63** (1991) 1126.
Capillary electrophoretic separations of proteins using nonionic surfactant coatings.
[29] G.J.M.Bruin, J.P.Chang, R.H.Kuhlman, K.Zegers, J.C.Kraak, H.Poppe, J.Chromatogr., **471** (1989) 429.
Capillary zone electrophoretic separations of proteins in polyethylene glycol-modified capillaries.
[30] S.A.Swedberg, Anal.Biochem., **185** (1990) 51.
Characterization of protein behavior in high-performance capillary electrophoresis using a novel capillary system.
[31] M.A.Moseley, L.J.Deterding, K.B.Tomer, J.W.Jorgenson, Anal.Chem., **63** (1991) 109.
Determination of bioactive peptides using capillary zone electrophoresis/mass spectrometry.
[32] J.Kohr, H.Engelhardt, Proceedings of the 13th Int.Symp. on Capillary Chromatography, Riva del Garda, Italy (1991) 1172.
Capillary electrophoresis with surface modified fused silica capillaries.
[33] J.K.Towns, F.E.Regnier, J.Chromatogr., **516** (1990) 69.
Polyethyleneimine-bonded phases in the separation of proteins by capillary electrophoresis.
[34] J.Wiktorowicz, J.Colburn, Electrophoresis (Weinheim), **11** (1990) 769.
Separation of cationic proteins via charge reversal in capillary electrophoresis.
[35] A.Wainright, J.Microcol.Sep., **2** (1990) 166.
Capillary electrophoresis applied to the analysis of pharmaceuticals.

8 Injection

8.1 Volume Restrictions for Injection

From the data in **Table 7.1**, it is clear that only extremely small volumes of sample can be applied. Considering the extra-column contribution of injection to peak dispersion :

$$\sigma_{tot}^2 = \sigma^2 + \sigma_{inj}^2 = \sigma^2 + \frac{l_{inj}^2}{12} \tag{8.1}$$

it can be calculated that, for a 50 cm column and a modest N = 100,000, and allowing a loss of 10% in plate count from injection, the length of the injected plug should be limited to less than 1.8 mm, which, in the case of a 50 μm ID column, is only 3.6 nl. In more efficient systems, with shorter and smaller columns, the allowed injection volume drops fast into the picoliter range with plug lengths of less than 1 mm. Microsyringes have been used [1] and rotary-type injector devices [2] have been designed, but most in combination with large diameter columns (> 200 μm ID) for which injection volume is less restricted. However, using a split-flow technique, nanoliter injections on 75 μm ID columns have been shown to be possible with a microsyringe and a piece of split vent capillary tubing [3]. In the great majority of cases, two injection techniques which are typical for capillary electrophoresis have been used : electrokinetic injection (or injection by electromigration) and hydrodynamic [1] injection. In both cases a plug-like injection profile is obtained by inserting the capillary in the sample solution and applying a force field for a controlled length of time, using an electric field or a pressure difference (**Figure 8.1**).

8.2 Quantitative Aspects of Injection

In both electromigration and hydrodynamic injection the injected quantity Q_{inj} in the plug with length l_{inj}, can be calculated from the injection time t_{inj} , the linear velocity of the injection plug v_{inj} and the concentration in the sample c_{inj} :

---

[1] *Sometimes referred to as 'hydrostatic' although injection is clearly related to moving liquids, certainly not at rest.*

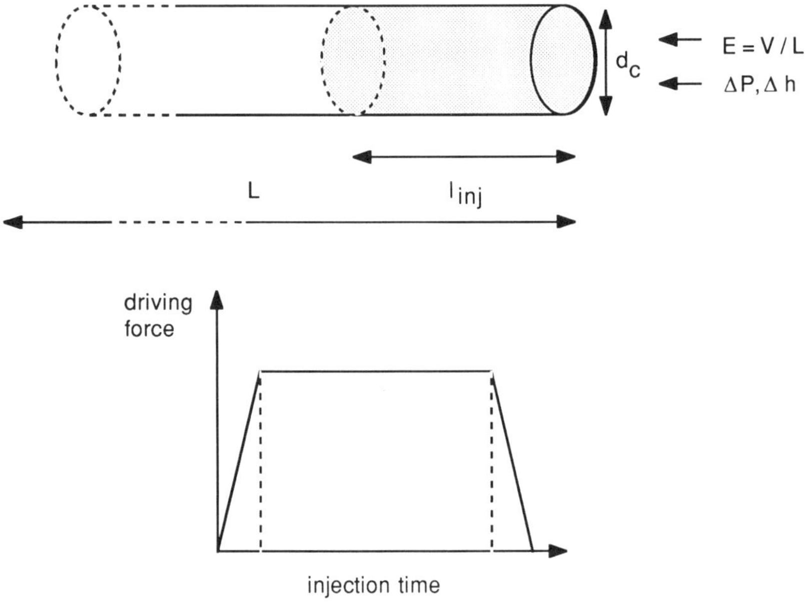

Figure 8.1
Plug-like injection by applying voltage or a pressure difference.

$$Q_{inj} = l_{inj} \cdot c_{inj} \cdot \frac{\pi\, d_c^2}{4} = v_{inj} \cdot t_{inj} \cdot c_{inj} \cdot \frac{\pi\, d_c^2}{4} \tag{8.2}$$

In the case of electromigration, v_{inj} is related to the mobility of the sample components, but also to the electroosmotic flow in the system, and the applied field during injection V_{inj} [4] :

$$v_{inj} = \frac{\mu \cdot V_{inj}}{L_V} = \frac{(\mu_{eo} + \mu_{ep}) \cdot V_{inj}}{L_V} \tag{8.3}$$

or

$$Q_{inj} = V_{inj} \cdot t_{inj} \cdot (\mu_{eo} + \mu_{ep})\ c_{inj} \cdot \frac{\pi\, d_c^2}{4 \cdot L_V} \tag{8.4}$$

As μ_{ep} will be different for the sample components, faster moving components will be proportionally oversampled, although this effect will be diminished when electroosmosis is much faster than the electrophoretic mobilities of the sample constituents [5]. Calculated Q_{inj} can be corrected by using the observed migration time t_M during analysis (with applied voltage V), provided that sample and run buffer are similar :

$$\mu = \mu_{eo} + \mu_{ep} = \frac{L_V \, . L_d}{t_M \, . V} \tag{8.5}$$

or

$$Q_{inj} = \frac{V_{inj} \, .t_{inj}}{V \, .t_M} \, . \; c_{inj} \, . \frac{\pi \, d_c^2 . L_d}{4} \tag{8.6}$$

The relative enrichment or bias factor b for 2 components [5] can be expressed as

$$b = \frac{\mu_{eo} + \mu_{ep,1}}{\mu_{eo} + \mu_{ep,2}} = \frac{t_{M,2}}{t_{M,1}} \tag{8.7}$$

so that

$$\frac{Q_{inj,1}}{Q_{inj,2}} = b \, . \frac{c_{inj,1}}{c_{inj,2}} \tag{8.8}$$

In eq.(8.5) to (8.8), it is assumed that components move in the same direction. With no or moderate electroosmosis, cations and anions will move in opposite direction. Sampling, instead of simply biased, will become selective for cations or anions. With appropriate control of the electroosmotic flow, even selective sampling of slow moving anions, against fast moving anions, can be considered.

For hydrodynamic injection, the linear velocity is determined by the Poiseuille-Hagen relation :

$$v_{inj} = \frac{\Delta P_{inj}}{8.L_V.\eta} \, . \frac{d_c^2}{4} \tag{8.9}$$

so that

$$Q_{inj} = \Delta P_{inj} \,.t_{inj} \,.c_{inj} \,. \frac{1}{\eta} \,. \frac{\pi.d_c^4}{128.L_V} \qquad (8.10)$$

In pneumatic injection, a pressure difference ΔP_{inj} is applied at the injection side (injection by pressure) or at the detector side (injection by suction or vacuum). For siphonic injection, the pressure difference is obtained by lifting the sample vial, creating a height difference Δh_{inj} between column in- and outlet. When using gravity, eq.(8.10) can be further elaborated to include the effect of Δh_{inj}. The pressure difference is then determined by the sample solution density ρ and Δh_{inj} :

$$\Delta P_{inj} = \rho \,. \Delta h_{inj} \,.g \qquad (8.11)$$

or

$$Q_{inj} = \Delta h_{inj} \,.t_{inj} \,.c_{inj} \,. \frac{g.\rho\,.}{\eta} \,. \frac{\pi.d_c^4}{128.L_V} \qquad (8.12)$$

No discrimination occurs between components and the injected aliquot is representative for the sample. This may seem to make hydrodynamic injection preferable over electrokinetic injection, but some practical limitations must be considered.

Assuming an elevation of 10 cm of the injection vial and using approximate values (10 cm $H_2O \approx 1000$ kg / m.sec^2 and $\eta_{buffer} \approx \eta_{water} \approx 0.001$ kg / m.sec) linear velocities during injection, plug lengths and injection volume per second of injection can be calculated. **Table 8.1** shows the limitations within which hydrodynamic injection can be used. If the column diameter is too large, resistance to flow is too low to be used reliably for injection. On the other hand, if the column diameter becomes too small, linear velocities are too low to be practical. Also, hydrodynamic injection cannot be used with gel-filled columns as the required pressure would disrupt the gel in the column.

The data in **Table 8.1** can also be used to estimate the importance of hydrodynamic flow caused by level differences between buffer vials. When electromigration injection is used with relatively wide columns, it is necessary to balance the liquid levels in both vials to avoid supplementary siphoning effects during injection. Similarly, level differences between buffer vials can cause hydrodynamic flow during analysis.

In electrokinetic and hydrodynamic injection, the length of the injection plug is determined by the product of the driving force and the application time (eq.(8.2)). Automation and electronic control of the injection time is decisive in obtaining good reproducibility [4,6-8], siphoning being more reproducible than electromigration [4,6]. Almost any application of the driving force will show some inertia (**Figure 8.1**) : lifting time of the vial, initial ramp when applying voltage, building up of pressure [4]. Therefore, for a constant product of force and time, injected quantities can differ. Apart from micro-processor controlled correction for this effect, it can also be minimized by selecting a relatively low force level with corresponding longer injection time [9].

Table 8.1 Approximate linear velocities and injected volumes for injection by siphoning (for Δh = 10 cm [a]), column length = 1 m [b]))

Column ID (µm)	mm/sec [c])	nl/sec
500	7.81	1534
100	0.313	2.454
75	0.176	0.777
50	0.078	0.153
25	0.020	0.010
10	0.003	0.0002

a) Multiply with Δh (cm) / 10 for other elevations
b) Divide by length in meter for other column lengths
c) For plug length, multiply by injection time (sec)

8.3 Influence of Injection Volume on Efficiency

The importance of limiting the injection plug length has already been mentioned and its contribution to plate height, according to eq.(8.1) has been considered by several authors [10-13]. More generally, σ_{inj}^2 can be written as :

$$\sigma_{inj}^2 = \frac{l_{inj}^2}{K_i^2} \tag{8.13}$$

with K_i^2 = 12 for a rectangular concentration profile. Several processes (diffusion, parabolic flow, voltage rise and fall time ...), related to the injection method, can distort this profile. For comparable peak areas, no difference in efficiency could be found between siphoning and electromigration injection methods, although it was acknowledged that, on further reducing other sources of band broadening, differences could become apparent [4].

Especially when performed manually, the importance of a good injection technique has been stressed [13,14]. As demonstrated by *Grushka* and *McCormick* [11], sample can enter the column during insertion of the column in the sample solution, while left idle there and also during withdrawal. This so-called extraneous injection can contribute to band broadening. A minimal injection length has been postulated [53], originating from mixing at the interface between capillary outlet and sample solution, and was estimated at 200 µm for 75 µm ID capillaries.

Sample can be adsorbed on the outside of the capillary and then slowly enter the column, after being diluted in the run buffer vial. Rinsing the capillary has been considered unnecessary as it would disturb the sample band [16]. But *Lux* et al. showed that, using an automated injection system, dipping the capillary in a buffer filled vial, before repositioning it in the run buffer vial, reduced peak tailing [17]. The quality of the cut at the capillary end also seems important. In their open-tubular capillary electrochromatographic experiments (see Chapter 4), also using electromigration injection, *Bruin* et al. found that cutting the end of the capillary obliquely increased σ_{inj}^2 more than two-fold [18].

Huang et al. [11] have concluded that in almost all reported CZE separations, peak profiles were dominated by the sample injection length. In relation to available integrated instrumentation, this statement needs some consideration, especially regarding pneumatic injection methods. In those instruments, pressure or vacuum, the level being a given quantity, is applied during a period, programmable by the user, generally ranging between 1 and several tens of seconds. According to our experience, this freedom of choice is illusory as, unless special precautions are taken that will be discussed in the next section, only the very lowest values can be used. **Figure 8.2** shows the analysis of organic anions at different injection times, showing a serious decrease in efficiency. The injection pressure on this instrument is 0.5 psi [19] or 35 cm H_2O [1] . Although peak areas increase linearly, as should be the case (**Figure 8.3a**), peak heights saturate quickly (**Figure 8.3b**); this can be

[1] *This pressure unit is used for easy comparison with siphonic injection. In the latter case, typical elevations are 5 to 10 cm.*

correlated with a deterioration of the efficiency (**Figure 8.3c**). **Figure 8.4** shows the same effects with another type of analysis and on another instrument (ΔP_{inj} unknown). For a third instrument it is known that the vacuum used, has a value of 5" Hg [20] or 172 cm H_2O. Using this real-life value, the relation with other instrumental parameters and separation efficiency can be estimated.

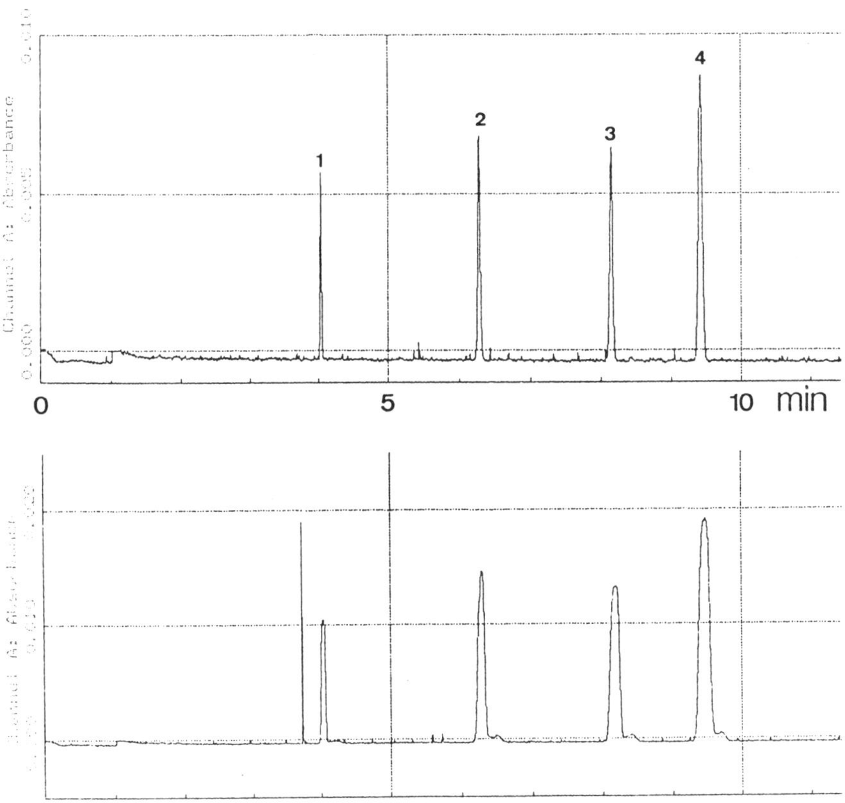

Figure 8.2
Pneumatic injection and the CZE analysis of organic anions. Capillary : 75 µm x 50/57 cm. Buffer : 12.5 mM borax, pH 9.2. Applied voltage : 20 kV (at 25°C). Detection : UV, 214 nm. Hydrodynamic injection : 2 sec (top), 8 sec (bottom). The sample is dissolved in the run buffer.
1 = Mesityloxide (t_0) , 2 = Phenylglycine, 3 = Mandelic acid, 4 = Benzoic acid.

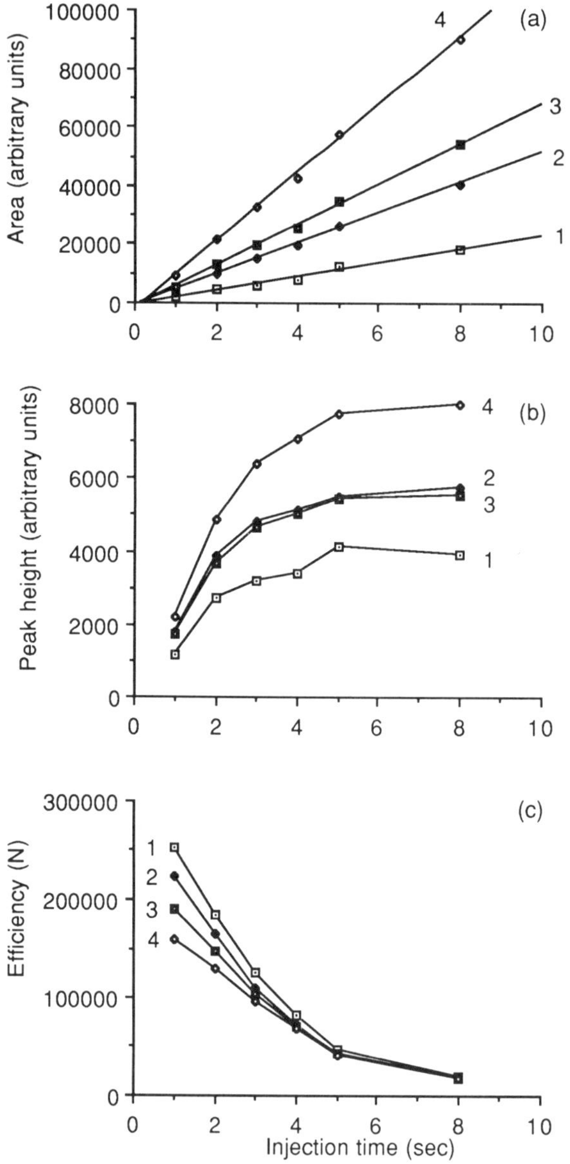

Figure 8.3
Peak area, height and efficiency as a function of injection time. Peak numbers and conditions as in **Figure 8.2**.

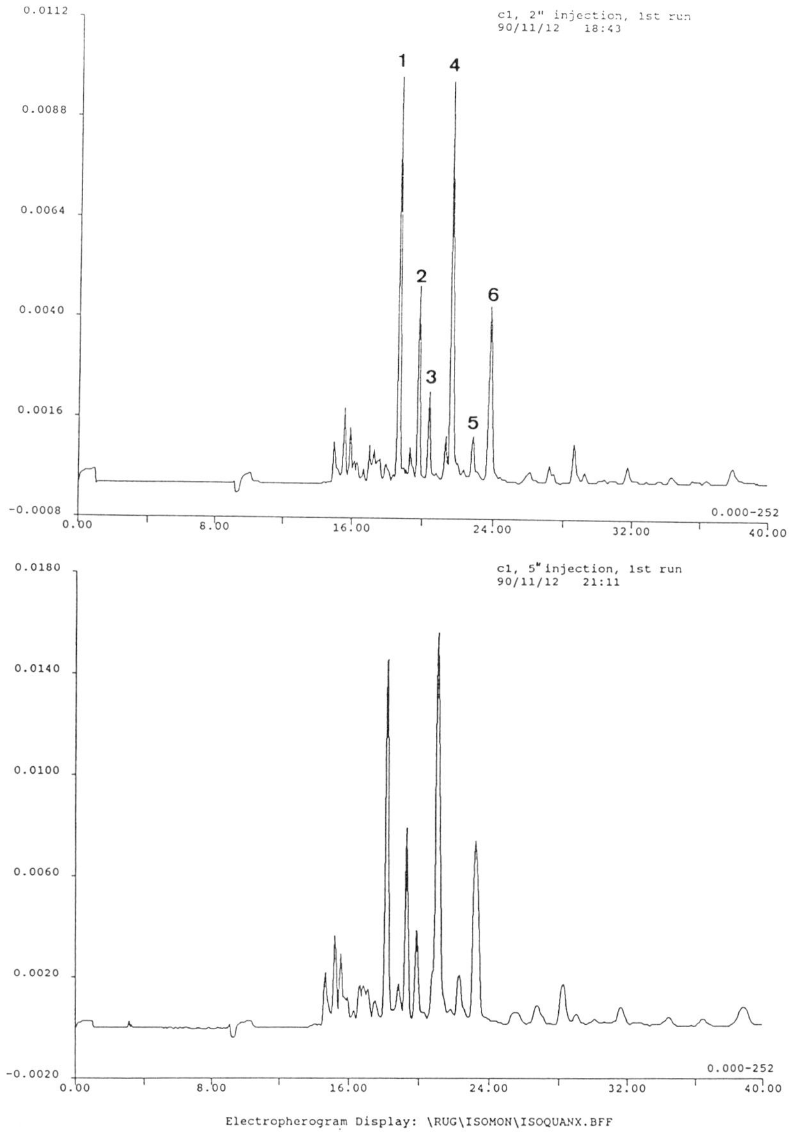

Figure 8.4

Pneumatic injection and the MEKC analysis of iso-α-acids. Capillary : 50 μm x 63/70 cm. Buffer : 30 mM Na_2HPO_4/HCl, pH 7.6, 40 mM SDS. Applied voltage : 20 kV (at 25°C). Detection : 252 nm. Hydrodynamic injection : 2 sec (top), 5 sec (bottom). The sample is dissolved in the run buffer.
1,3,4 : Cis-iso-humulones, 2,5,6 : Trans-iso-humulones.

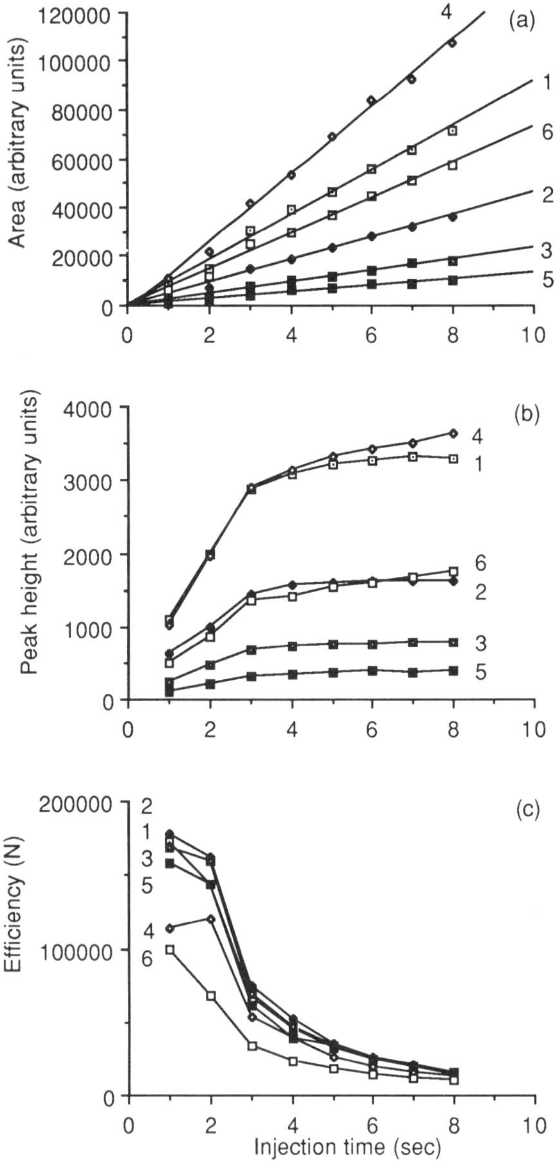

Figure 8.5
Peak area, height and efficiency as a function of injection time. Peak numbers and conditions as in **Figure 8.4**.

Assuming that the buffer viscosity is equal to that of water (0.001 kg / m.sec), and using eq.(8.9) (or **Table 8.1**), the injection plug length per second can be expressed as

$$l_{inj} = \frac{f(r_c)}{L} t_{inj} \tag{8.14}$$

where $f(r_c) = 1.34 \times 10^{-3}$, 3.03×10^{-3} and 5.40×10^{-3} m^2/sec for 50, 75 and 100 μm ID columns. Efficiency can now be represented by :

$$N = \frac{L^2}{2 D_m \frac{L}{v} + \frac{1}{12}\left(\frac{f(r_c)\, t_{inj}}{L}\right)^2} \tag{8.15}$$

Figure 8.6 shows calculated efficiencies as a function of the injection time, assuming a 50 cm column and a linear velocity of 1 mm/sec (analysis time = 8.3 min). The theoretically maximal attainable efficiencies are 250,000 (high D_m, common organic ions) and 2.5×10^6 (low D_m, biopolymers). Although only one of the non-diffusional causes of band broadening is considered, efficiency is already much lower than the theoretical value. The need to limit injection time is obvious. Larger diameter columns are more affected as they show less flow resistance during injection. The profile for small molecules is remarkably similar to those observed in **Figure 8.3c** and **Figure 8.5c**.

When the injection time is held constant at 1 sec, the influence of applied voltage, or linear velocity (**Figure 8.7**), is much less than would be predicted from ideal behaviour (i.e. $l_{inj} = 0$). Especially for large ions or with large column diameters, the effect of increased voltage levels off fast.

Rewriting eq.(8.15) as

$$N = \frac{L^2}{2 D_m \frac{L^2}{\mu V} + \frac{1}{12}\left(\frac{f(r_c)\, t_{inj}}{L}\right)^2} \tag{8.16}$$

allows calculation of the effect of column length when the applied voltage is held constant (**Figure 8.8**). Analysis times range from 5.3 min at 40 cm to over 33 min at 100 cm. Using sufficiently long, small diameter columns offers the best chance of approaching theoretically possible efficiency values. However, if high efficiency is

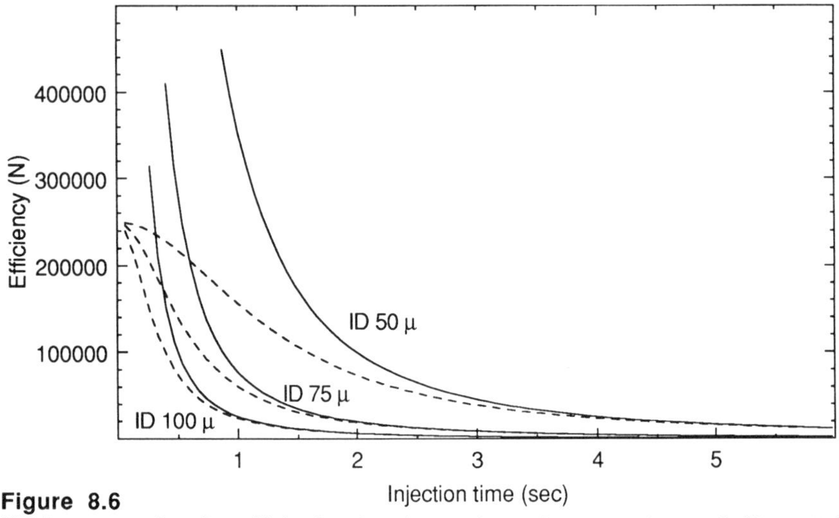

Figure 8.6
Efficiency as a function of injection time for 3 column diameters, for small ($D_m = 10^{-9}$ m^2 / sec, broken lines) and for large ions ($D_m = 10^{-10}$ m^2 / sec, solid lines). According to eq.(8.15) (L = 0.5 m, v = 1 mm / sec).

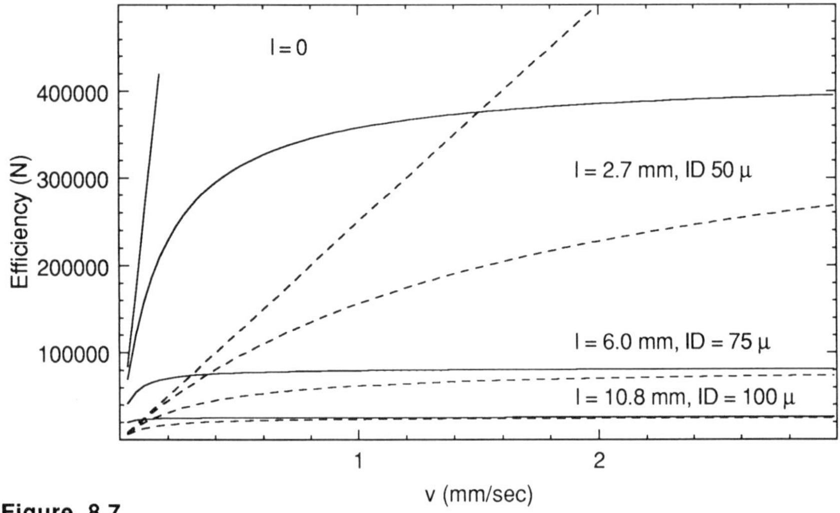

Figure 8.7
Efficiency as a function of linear velocity for 3 column diameters, compared with the theoretical maximum (l = 0) for small ($D_m = 10^{-9}$ m^2 / sec, broken lines) and for large ions ($D_m = 10^{-10}$ m^2 / sec, solid lines). According to eq.(8.15) (L = 0.5 m, t_{inj} = 1 sec, l is the length of the injection plug).

required reducing analysis time by using short columns is not possible, unless injection methods are adapted.

Alternatively, column length can be increased while the linear velocity is held constant by increasing the applied voltage simultaneously (**Figure 8.9**). This seems more effective in obtaining high separation efficiency, as two factors are now cooperating, provided that the necessary field does not cause excessive heating.

Summarizing, it is obvious that the selection of column length for a particular analysis should not be based solely on the ideal conditions described in Chapter 3. Short columns can only be used when injection procedures are adapted accordingly. Similarly, a small column radius is important not only for reducing thermal gradients, but also for its flow resistance during hydrodynamic injection when the pressure cannot be freely selected. The calculated examples are related to CZE because the theoretical value of the efficiency is easily modelled using diffusion coefficients. However, it is obvious from **Figure 8.4** that comparable conclusions are valid in MEKC.

8.4 Sample Matrix Effects and Sample Stacking

With electromigration injection, it has been observed that the injected quantity varies with the sample conductivity. On using a 0.02 mM phosphate solution as a sample, and adding 0.02-1.00 mM chloride, peak areas for the phosphate were linearly correlated with the sample resistance [21]. Similar results were obtained for cations in solutions of varying buffer concentration [5]. This effect has been ascribed to the higher ion mobility and electroosmotic flow at lower ionic strength [5] but also to the increased local field strength at increased resistance [21,22].

The adverse effect of the initial zone width can often be counteracted by focusing phenomena during or shortly after injection. Peak sharpening is observed when the sample has a lower conductivity than the buffer on dissolving it in water or diluted buffer [23-25]. Using gravity injection, *Burgi* et al. could increase injection time ten-fold while maintaining the same peak width, resulting in a ten-fold increase in detectability [25]. The phenomenon is related to the zone boundary sharpening effect in isotachophoresis and can be further enhanced by the presence of very mobile ions in the sample that can act as a leading electrolyte [26]. The presence of chloride (from salt in the sample) has been shown to have a strong sharpening effect in the analysis of anions [27]. Consistent application of this principle leads

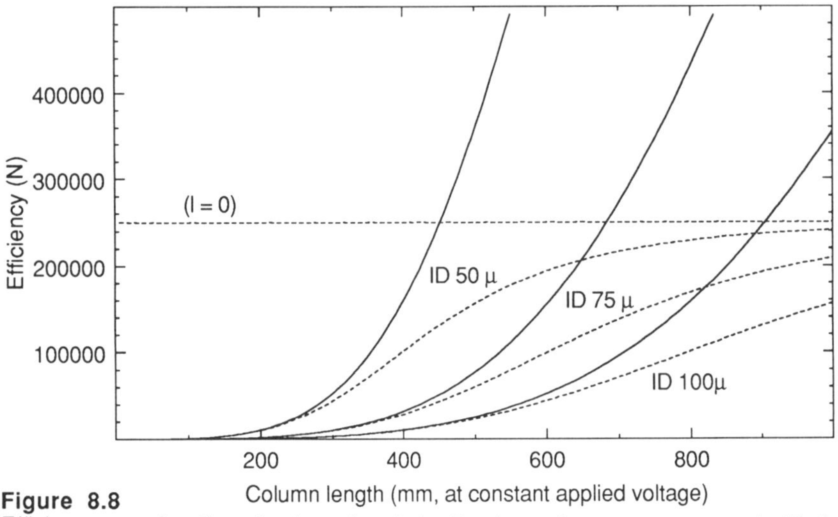

Figure 8.8
Efficiency as a function of column length for 3 column diameters, compared with the theoretical maximum (l = 0, low D_m not shown) for small ($D_m = 10^{-9}$ m^2/sec, broken lines) and for large ions ($D_m = 10^{-10}$ m^2/sec, solid lines). According to eq.(8.16) (t_{inj} = 1 sec, $\mu V = 5 \times 10^{-4}$ m^2 / sec).

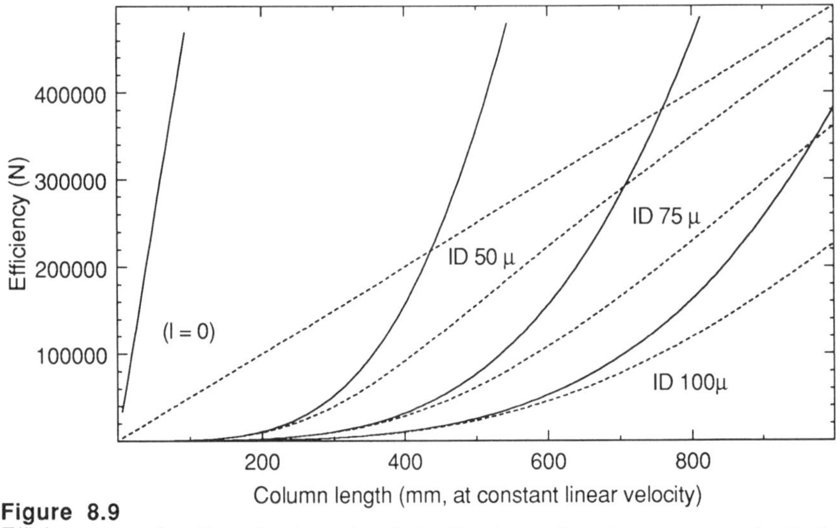

Figure 8.9
Efficiency as a function of column length for 3 column diameters, compared with the theoretical maximum (l = 0) for small ($D_m = 10^{-9}$ m^2 / sec, broken lines) and for large ions ($D_m = 10^{-10}$ m^2 / sec, solid lines). According to eq.(8.15) (t_{inj} = 1 sec, v = 1 mm / sec).

eventually to a hyphenated technique, ITP-CZE, in which isotachophoretic sample concentration is followed by zone electrophoretic separation, allowing the analysis of large, diluted samples [28-31]. In another focusing technique, a diluted peptide solution at high pH was sandwiched between two low pH zones [32]. The amphoteric nature of the peptides results in opposed velocities in the two neighbouring zones and focusing at the zone boundary.

The sample preconcentration techniques that have just been mentioned all rely on the electrophoretic mobility of the sample constituents; their nature precludes their simple transfer to MEKC. Reports of sample introduction effects specific for MEKC, and their application to preconcentration are exceptional. Chromatographic sample stacking, i.e. focusing through in-column micellization, apart from being less effective through the movement of the pseudo-stationary phase, suffers from contradictory conditions. It will be most important for strongly micellized compounds which will be insoluble if the surfactant is absent in the sample solution. On the other hand, using organic modifiers instead of surfactant for sample solubilization, causes peak shape deterioration [9].

Sample stacking, similar to that in CZE, can only be expected with ionic compounds (**Figure 8.10**) and has been observed after elimination of the surfactant from the sample solution for anionic porphyrins [33] and iso-α-acids [34]. Peak compression was also observed for those compounds with urine [33] and beer [34] samples respectively.

8.5 Special Injection Techniques

In the case of microsequencing of proteins, one sometimes has to deal with sample volumes in the microliter range. There is also a growing interest in the properties of very small, individual structures : analysis of the cytoplasm of single nerve cells [35-37], or the analysis of single rain droplets [38]. Decreasing the column inner diameter allows analysis of very small aliquots, provided that a sensitive detector is available. However, a certain amount of sample volume overhead remains necessary to introduce column and electrode during injection. This overhead is not reduced proportionally with the volume actually sampled, as even for very small internal diameter columns the outer diameter remains generally above 100 μm.

Most commercial instruments are now (optionally) equipped with microvials, which allow nanoliter-sampling from a few microliter, using conventional columns

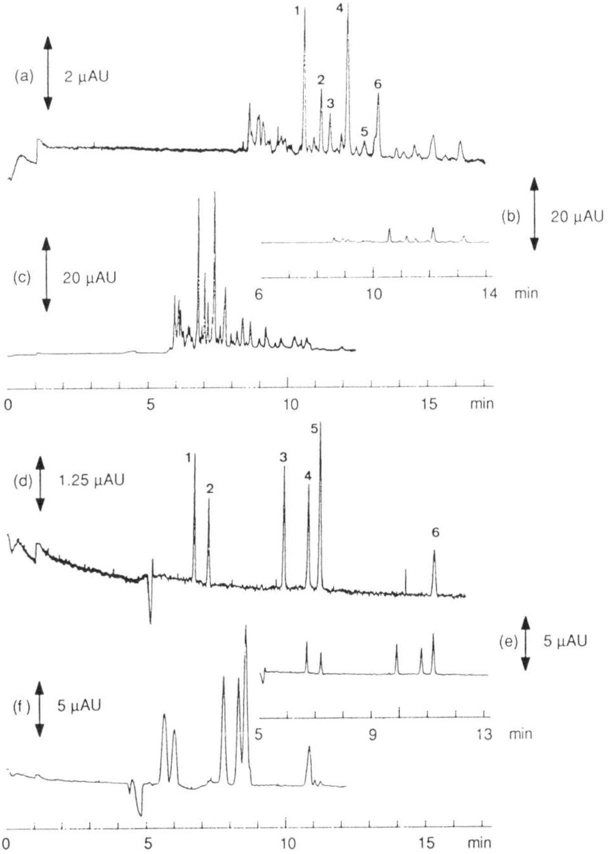

Figure 8.10
Sample stacking in MEKC : ionic versus nonionic solutes [34].

(a) + (b) : Iso-α-acid salts (in a preisomerized hop extract, 0.1 mg/ml) dissolved in buffer, injection time 2 sec (1,3,4 : Cis-isohumulones, 2,5,6 : Trans-isohumulones)

(c) : Iso-α-acid salts dissolved at the same concentration in diluted (10x) buffer, injection time 20 sec

(d) + (e) : Nonionic solutes dissolved in buffer, injection time 2 sec (1= Aniline, 2 = Benzyl alcohol, 3 = Nitrobenzene, 4 = Benzyl cyanide, 5 = Acetophenone, 6 = Toluene, each approximately 0.1% v/v)

(f) Nonionic solutes dissolved at the same concentration in diluted (10x) buffer, injection time 20 sec

Column : 75 μm x 50/57 cm. Buffer : 40 mM SDS in 30 mM phosphate (pH 7.5). Applied voltage : 20 kV. Detection : UV, 254 (a-c) or 200 (d-f) nm. Pneumatic injection.

(50 or 75 μm ID). *Yin* et al. eliminated the need to insert an electrode by machining a titanium electrode into the shape of a sample vial, claiming to be able to handle 1-2 μl sample droplets [39].

For the analysis of nerve cell cytoplasma, a dual-barreled microinjector, containing a carbon fiber electrode, has been designed with a tip diameter of less than 50 μm [37]. By elimination of the electrode, placing it in the solution surrounding the test animal (a pond snail), and using etched columns, an injector with a 10 μm tip was constructed, allowing sub-picoliter sampling from a 200 μm diameter cell [36].

For high-speed analysis, using narrow, short columns (< 4 cm), the on-column sample gating technique has been developed [40]. A stream of fluorescently labeled sample is continuously photodegraded with the aid of a laser. By temporarily blocking the laser, a small plug of tagged, detectable material is allowed to enter the column. Injection times of 10 to 20 ms can be realized in this way.

8.6 Concentration Overloading and Electromigration Dispersion

Besides the volume, sample concentration can also affect efficiency. When the conductivity of the zone differs considerably from the background buffer, isotachophoretic effects can lead to zone deformation : electromigration dispersion [23].

For reasons of electroneutrality, sample ions displace part of the carrier electrolyte in the sample zone. When their mobility is different, local conductivity, and consequently resistance, will be changed by this displacement. As the current density must be constant, it follows from Ohm's law that the local field strength will also be changed. Let us consider the case when the carrier electrolyte has a higher mobility than the sample ions (**Figure 8.11**). When a sample ion diffuses out of the zone on the front side (relative to the migration direction), it will experience an electric field that is relatively low, be slowed down and overtaken again by the faster moving sample zone. This has a sharpening effect, related to similar processes in isotachophoresis, the background buffer functioning as the leading electrolyte (highest mobility). On the other hand, when an ion diffuses out of the sample zone on the rear side, it will also be slowed down and lag behind the center of the sample zone, more than it would on diffusion alone. The result is a tailed peak. Following similar reasoning, the leading profile of the peak can be explained when sample ions have a higher mobility than the background.

Mikkers et al. [23] have evaluated the conditions at which electromigration and diffusional dispersion are of the same order of magnitude :

$$\frac{c_{sample}}{c_{carrier}} \approx \frac{10 \,.\, D_m}{v_{sample} \cdot l_{inj} \cdot f\left(\frac{\mu_{sample}}{\mu_{carrier}}\right)} \qquad (8.17)$$

and concluded that for small molecules ($D_m = 10^{-9}$ m^2/sec), sample concentration should preferably be less than 1% of the carrier in order to reduce the effects of electromigration dispersion.

Examples can be found were all peak profiles can be observed in a single electropherogram [41,42]. Early eluting ions, with a mobility faster than the background, show peaks with a leading profile while later eluting, slowly moving ions show tailed peaks; in between there is a transition zone with symmetric peaks.

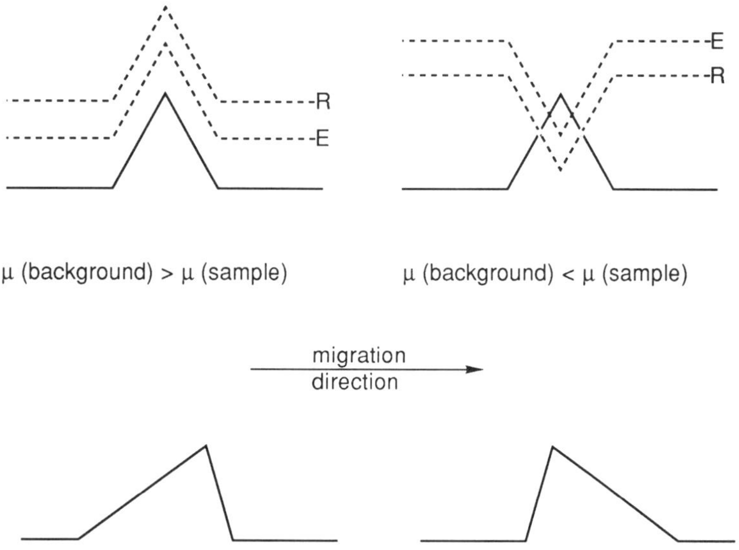

Figure 8.11
Electromigration dispersion. Changes in resistance and electric field along the sample zone and the resulting zone deformation.

References

[1] J.Sudor, Z.Stránsky, J.Pospíchal, M.Deml, P.Boček, Electrophoresis (Weinheim), **10** (1989) 802.
Step change of counterion - a new option in capillary zone electrophoresis.

[2] T.Tsuda, T.Mizuno, J.Akiyama, Anal.Chem., **59** (1987) 799.
Rotary-type injector for capillary zone electrophoresis.

[3] J.Tehrani, R.Macomber, L.Day, J.High Resolut.Chromatogr., **14** (1991) 10.
Capillary electrophoresis : an integrated system with a unique split-flow sample introduction mechanism.

[4] D.J.Rose, J.W.Jorgenson, Anal.Chem., **60** (1988) 642.
Characterization and automation of sample introduction methods for capillary zone electrophoresis.

[5] X.Huang, M.J.Gordon, R.N.Zare, Anal.Chem., **60** (1988) 375.
Bias in quantitative capillary zone electrophoresis caused by electrokinetic sample injection.

[6] S.Honda, S.Iwase, S.Fujiwara, J.Chromatogr., **404** (1987) 313.
Evaluation of an automatic siphonic sampler for capillary zone electrophoresis.

[7] S.Fanali, L.Ossicini, F.Foret, P.Boček, J.Microcol.Sep., **1** (1989) 190.
Resolution of optical isomers by capillary zone electrophoresis : Study of enantiomeric and diastereoisomeric Co(III) complexes with ethylenediamine and amino acid ligands.

[8] H.E.Schwartz, M.Melera, R.G.Brownlee, J.Chromatogr., **480** (1989) 129.
Performance of an automated injection and replenishment system for capillary electrophoresis.

[9] D.E.Burton, M.J.Sepaniak, M.P.Maskarinec, Chromatographia, **21** (1986) 583.
The effect of injection procedures on efficiency in micellar electrokinetic capillary chromatography.

[10] F.Foret, M.Deml, P.Boček, J.Chromatogr., **452** (1988) 601.
Capillary zone electrophoresis. Quantitative study of the effects of some dispersive processes on the separation efficiency.

[11] X.Huang, W.F.Coleman, R.N.Zare, J.Chromatogr., **480** (1989) 95.
Analysis of factors causing peak broadening in capillary zone electrophoresis.

[12] K.Otsuka, S.Terabe, J.Chromatogr., **480** (1989) 91.
Extra-column effects in high-performance capillary electrophoresis.

[13] H.K.Jones, N.T.Nguyen, R.D.Smith, J.Chromatogr., **504** (1990) 1.
Variance contributions to band spread in capillary zone electrophoresis.

[14] R.D.Smith, H.R.Udseth, J.A.Loo, B.W.Wright, G.A.Ross, Talanta, **36** (1989) 161.
Sample introduction and separation in capillary electrophoresis, and combination with mass spectrometric detection.

[15] E.Grushka, R.M.McCormick, J.Chromatogr., **471** (1989) 421.
Zone broadening due to sample injection in capillary zone electrophoresis.

[16] K.D.Lukacs, J.W.Jorgenson, J.High Resolut.Chromatogr.Chromatogr.Commun., **8** (1985) 407.
Capillary zone electrophoresis : Effect of physical parameters on separation efficiency and quantitation.

[17] J.A.Lux, H.-F.Yin, G.Schomburg, Chromatographia, **30** (1990) 7.
Construction, evaluation and analytical operation of a modular capillary electrophoresis instrument.

[18] G.J.M.Bruin, P.P.H.Tock, J.C.Kraak, H.Poppe, J.Chromatogr., **517** (1990) 557.
Electrically driven open-tubular liquid chromatography.

[19] T.Satow, A.Machida, K.Funakushi, R.Palmieri, J.High Resolut.Chromatogr., **14** (1991) 276.
Effects of the sample matrix on the separation of peptides by high performance capillary electrophoresis.

[20] S.E.Moring, J.C.Colburn, P.D.Grossman, H.H.Lauer, LC*GC, **8**(1) (1990) 34 (also : LC*GC Int., **3**(2) (1990) 46).
Analytical aspects of an automated capillary electrophoresis system.

[21] L.Gross, E.S.Yeung, J.Chromatogr., **480** (1989) 169.
Indirect fluorimetric detection and quantification in capillary zone electrophoresis of inorganic anions and nucleotides.

[22] X.Huang, J.Ohms, J.Chromatogr., **516** (1990) 233.
Non-uniform electrical field effect caused by different concentration of electrolyte in capillary zone electrophoresis.

[23] F.E.P.Mikkers, F.M.Everaerts, Th.P.E.M.Verheggen, J.Chromatogr., **169** (1979) 1.

Concentration distributions in free zone electrophoresis.
[24] F.E.P.Mikkers, F.M.Everaerts, Th.P.E.M.Verheggen, J.Chromatogr., **169** (1979) 11.
High-performance zone electrophoresis.
[25] D.S.Burgi, R.-L.Chien, J.Microcol.Sep., **3** (1991) 199.
Application of sample stacking to gravity injection in capillary electrophoresis.
[26] J.L.Beckers, F.M.Everaerts, J.Chromatogr., **508** (1990) 19.
Isotachophoresis with two leading ions and migration behaviour in capillary zone electrophoresis. II. Migration behaviour in capillary zone electrophoresis.
[27] Th.P.E.M.Verheggen, A.C.Schoots, F.M.Everaerts, J.Chromatogr., **503** (1990) 245.
Feasibility of capillary zone electrophoresis with suppression of electroendosmotic flow in completely closed systems.
[28] V.Dolnik, K.A.Cobb, M.Novotny, J.Microcol.Sep., **2** (1990) 127.
Capillary zone electrophoresis of dilute samples with isotachophoretic preconcentration.
[29] F.Foret, V.Sustacek, P.Boček, J.Microcol.Sep., **2** (1990) 229.
On-line isotachophoretic sample preconcentration for enhancement of zone detectability in capillary zone electrophoresis.
[30] D.Kaniansky, J.Marák, J.Chromatogr., **498** (1990) 191.
On-line coupling of capillary isotachophoresis with capillary zone electrophoresis.
[31] D.S.Stegehuis, H.Irth, U.R.Tjaden, J.van der Greef, J.Chromatogr., **538** (1991) 393.
Isotachophoresis as an on-line concentration pretreatment technique in capillary electrophoresis.
[32] R.Aebersold, H.D.Morrison, J.Chromatogr., **516** (1990) 79.
Analysis of dilute peptide samples by capillary zone electrophoresis.
[33] R.Weinberger, E.Sapp, S.Moring, J.Chromatogr., **516** (1990) 271.
Capillary electrophoresis of urinary porphyrins with absorption and fluorescence detection.
[34] R.Szücs, J.Vindevogel, P.Sandra, unpublished results.
[35] R.T.Kennedy, M.D.Oates, B.R.Cooper, B.Nickerson, J.W.Jorgenson, Science, **246** (1989) 57.
Microcolumn separations and the analysis of single cells.
[36] A.G.Ewing, T.M.Olefirowicz, Anal.Chem., **62** (1990) 1872.
Capillary electrophoresis in 2 and 5 μm diameter capillaries : application to cytoplasmatic analysis.
[37] R.A.Wallingford, A.G.Ewing, Anal.Chem., **59** (1987) 678.
Characterization of a microinjector for capillary zone electrophoresis.
[38] K.Bachmann, K.-H.Steeg, T.Groh, A.Roder, I.Haag, I.Haumann, S.Engelmann, Proceedings of the 13th Int.Symp. on Capillary Chromatography, Riva del Garda, Italy (1991) 799.
The use of Micro-HPLC and capillary zone electrophoresis for the analysis of individual atmospheric rain droplets.
[39] H.-F.Yin, S.R.Motsch, J.A.Lux, G.Schomburg, J.High Resolut.Chromatogr., **14** (1991) 282.
A miniature device for electrokinetic or hydrodynamic sample introduction from small volumes in capillary electrophoresis.
[40] C.A.Monnig, J.W.Jorgenson, Anal.Chem., **63** (1991) 802.
On-column sample gating for high-speed capillary zone electrophoresis.
[41] F.Foret, S.Fanali, L.Ossicini, P.Boček, J.Chromatogr., **470** (1989) 299.
Indirect photometric detection in capillary zone electrophoresis.
[42] W.R.Jones, P.Jandik, Int.Lab.,1991(may), 61.
New methods for chromatographic separations of anions.

9 Detection

9.1 Detection Methods

Optical detection methods, especially UV absorbance and fluorescence, have been extensively used and until now the only ones that have been implemented in commercial instruments. Therefore, the following discussion is limited to optical methods. Literature references to other detection methods can be found in **Table 9.1**. Indirect detection methods are described in a separate paragraph.

Table 9.1 Non-optical detection methods, used in capillary electrophoretic techniques.

Electrodetection	
Potential gradient a)	[1-3]
Conductivity	[4-11]
Amperometry b)	[11-20]
Potentiometry	[21,22]
Mass spectrometry c)	
Ion spray	[23,24]
Electrospray d)	[25-35]
Coaxial CF-FAB e)	[36-41]
Radio-activity	[42-44]
Ion Mobility	[45]

a) Obsolete now, almost exclusively when CZE is performed in existing isotachophoretic apparatus.
b) For a discussion of the compatibility with surfactant containing buffers (MEKC) see [15].
c) Viewed from a different vantage point, CZE can be considered as a sample introduction method for the mass spectrometer.
d) Mass spectra of surfactants have been obtained (MEKC) [28] but applications to analytes have not yet been reported.
e) CF-FAB = Continuous-Flow Fast Atom Bombardment

9.2 On-Column Detection

The volume restrictions that apply to injection are equally valid for detection. Detectors must be miniaturized according to the column dimensions (**Table 7.1**) : a detection volume at the nanoliter, and even subnanoliter, level is required. In view of this restriction, detachable detection cells serving as part of a stand-alone detector are no longer usable. In fact, a small section of the column itself is used as the detection volume : on-column detection [46,47]. Turbulence related to cell geometry or originating in connectors is avoided and cell volume is related to the column diameter. Consequently, the remaining relevant parameter of the detection cell is the length used for illumination, from which the contribution to zone dispersion can be estimated [48-50] :

$$\sigma_{det}^2 = \frac{l_{det}^2}{12} \qquad (9.1)$$

Although the tolerance is dependent on column length and other zone dispersion factors, calculations by *Otsuka* and *Terabe* [49] based on eq.(9.1), suggest that the detection cell length should be less than 1 mm. Using a different approach, *Wang* et al. [51] come to similar conclusions, stating that peak distortion becomes significant when l_{det} becomes larger than 1 peak standard deviation, so that preferably

$$l_{det} < \frac{L}{\sqrt{N}} \qquad (9.2)$$

which yields a limiting value of 0.9 mm for L = 50 cm and N = 300,000.

On-column detection requires that part of the outer coating of the fused silica column is removed to create a small, transparent window. Use of a flame is possible but results in rather wide windows. Although the effectively illuminated part is restricted by a slit or pinhole, the risk of breaking the column is increased. Several devices have been described which make better size-controlled windows. *Lux* et al. use a heated filament to make defined windows [52]. *McCormick* and *Zagursky* have constructed a mechanical stripper that can be used after modification of the interior column with a heat-sensitive material [53]. The windows are also free of charred flakes of burned polyimide.

9.3 UV Absorption Detection

Exceptionally, dedicated UV absorbance detectors have been built for on-column use [54,55], but more often LC detectors have been modified for that purpose. On-column UV absorbance detection poses two main problems. First, the path length is very short, less than 100 μm, and secondly, unlike the plane-parallel arrangement in conventional detection cells, the entry and exit borders are curved. Through refractive index effects, the illuminated column will act as a lens [56].

Figure 9.1 shows light paths through tubular shaped cells of various diameters and various illumination widths. The ray tracings were calculated in relation to miniaturized liquid chromatography with packed columns [56] for which the cell diameters are less restricted (100-320 μm ID as in **Figure 9.1**). However, the conclusions are essentially the same for the smaller columns in capillary electrokinetic techniques, although sometimes more difficult to implement.

At about one to two cell diameters from the cell center, the emerging light bundle is dispersive, and the degree of dispersion is a function of the refractive index of the cell interior. The refractive index can change as a result of solvent composition (although gradients are less common in CZE), the presence of analytes or temperature variations. When the photocell is placed too far away, only a fraction of the light is captured, and that fraction will be a function of the refractive index of the cell contents. In other words, even when no light is absorbed, the actual amount of light measured (the absorbance) can vary. Therefore the photocell must be positioned close to the cell or a refocusing device must be used. Alternatively, optical fibers can be used to guide the light to the photocell [5,57,58].

Light that strikes the rim of the cell is excessively dispersed (**Figure 9.1a**). With wider cells this can easily be avoided (see e.g. **Figure 9.1b** vs. **c**). With smaller cells, 50 μm or less, reduction of the illumination width becomes more difficult and also strongly reduces the amount of light that reaches the detector, resulting in increased noise [54]. Here again focusing techniques can help to increase the amount of light that passes through the cell. In some commercial instruments, ball lenses are used for that purpose [59]. For focusing light over a longer illumination length (using a 50 μm x 0.8 mm slit) a cylindrical plano-convex lens has been used, resulting in a 10-fold increase in sensitivity [60]. Although demonstrated in miniaturized chromatography, this detection cell length would be tolerable in many CZE applications [61].

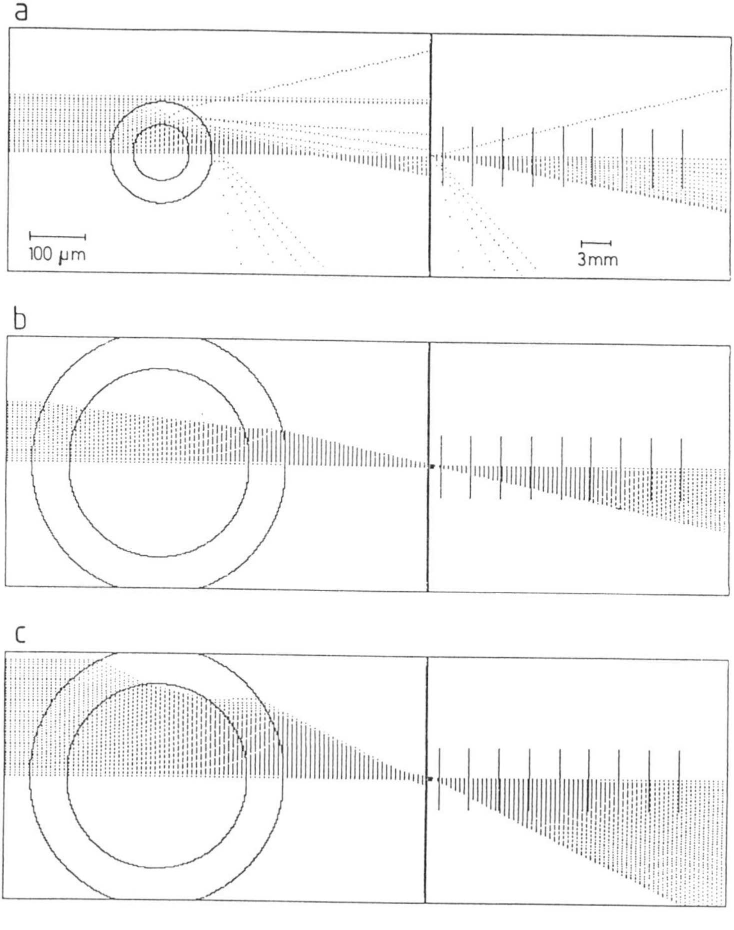

Figure 9.1
Left side : Light paths through tubular cells (cross section perpendicular to the column).
Right side : 6 mm wide photocells positioned at distances ranging from 1 to 25 mm.
(a) 100 µm ID, 180 µm OD, beam width 200 µm
(b) 320 µm ID, 450 µm OD, beam width 200 µm
(c) 320 µm ID, 450 µm OD, beam width 400 µm
(Repinted with permission [56] , © Hüthig Verlag)

Provided that refractive index effects are well controlled, reasonable linearity can be obtained. A linear range between 0.001% and 0.1% has been reported for proteins [55]. A wider range has been considered of less importance as higher solute concentration would lead to zone broadening (by electromigration dispersion) anyway [55].

The main problem remains at the lower concentration side. With current technology, limits of detection are in the 10^{-4}-10^{-7} M range (i.e. ppm- and sub-ppm-level). Values reported for lysozyme (MW = 14,000), 10^{-7} M at 229 nm [54], 10^{-8} M at 193 nm [55], are probably not representative of low-molecular weight organic chemicals, as with biopolymers the UV detector tends to be mass-sensitive rather than concentration-sensitive [55].

Attempts have been made to increase the path length to obtain better sensitivities. Rectangular borosilicate capillaries are commercially available with a short side of 50 μm and long sides up to 1000 μm. By illumination across their long axis, the effective path length is increased [62]. With long sides up to 340 μm, resolution is not affected but the gain in sensitivy is limited. When increased to 1000 μm, efficiency is reduced but sensitivity is increased 15-fold. However, those capillaries are less flexible, not protected by an external polyimide-coating, and very fragile.

Using fused silica bending technology, developed for the construction of 2 cm path length cells for micro-LC [63], *Chervet* et al. [64] have constructed a 3 mm Z-shaped on-column cell (ID 75 μm), resulting in a 5 to 6-fold increase in sensitivity. The loss of efficiency, estimated at 30% (depending upon column length) could be acceptable in some applications.

Tolerating the same level of efficiency loss, *Grant* and *Steuer* [65] used a more complicated method to obtain effective path lengths of 3 mm. The column is axially illuminated at the end and the absorbance is measured indirectly, at right angles to the light path, through the fluorescence of a tracer compound, added to the buffer.

On-column detection can also be used with diode array [66-68] or rapid spectral scanning detectors [69] (**Figure 9.2**), allowing on-line recording of spectra for peak identification.

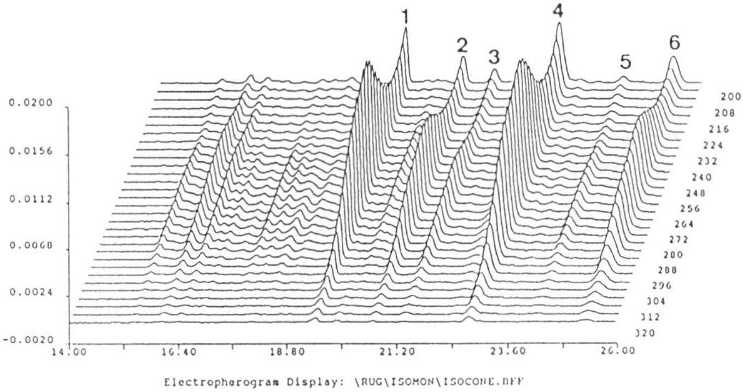

Figure 9.2
MEKC analysis of iso-α-acids with multi-wavelength UV detection. Capillary : 50μm x 63/70 cm. Buffer : 30 mM Na_2HPO_4/HCl, pH 7.6, 40 mM SDS. Applied voltage 20 kV (at 20°C). *(Reprinted with permission [69] , © Hüthig Verlag)*

9.4 Fluorescence Detection

The on-column method can also be used for fluorescence detection. Dedicated detectors have been built [70] or existing detectors have been modified (e.g. [50,71]). In the latter case, lens systems are generally present that allow efficient collection of the excitation light. *Kurosu* et al. [72] reduced light scattering by immersing the on-column cell in a liquid with a refractive index similar to that of fused silica, so that the cell boundary becomes effectively flat (if the liquid is contained in a square fused silica cell).

Although the comparison is biased by absorbance and fluorescence characteristics, fluorescence detection is considered more sensitive than absorbance, a factor of 15 being reported for derivatized amino acids [73]. However, with conventional light sources (deuterium, tungsten or xenon arc lamps) the gain is rather limited. With lasers, more light can be focused into small capillaries, which has resulted in impressive mass detection limits. The detection of 1,100 molecules of tagged arginine has been reported [74]. In some cases however, low mass detection limits are mainly due to detection volume reduction, and are not reflected in equivalently low concentration detection limits. Some results obtained with laser-induced fluorescence are summarized in **Table 9.2**.

Table 9.2 Detection limits obtained with laser induced fluorescence.

Compounds	Fluoro-phore a)	Mode b)	d_c (µm)	Laser c)/ λ_{em} (nm)	MDL d) (moles)	CDL e) (moles/l)	
Amino acids	DNS	CZE	75	H/325	5×10^{-17}	10^{-7}	[75]
Methotrexate	-	CZE	75	H/325	-	3×10^{-9}	[76]
Methotrexate	-	CZE	200	H/325	-	5×10^{-10}	[76]
Amino acids	FITC	CZE	50 f)	A/488	$<1.5 \times 10^{-19}$	$<10^{-10}$	[77]
Amino acids	NDA	CZE	10	H/442	$<2.5 \times 10^{-18}$	$<4 \times 10^{-8}$	[78]
Amino acids	OPA	CZE	75	H/325	1.4×10^{-14}	10^{-6}	[79]
Amino acids	FITC	CZE	50 f)	A/488	$<6 \times 10^{-21}$	$<6 \times 10^{-12}$	[74]
Myoglobin	OPA	CZE	50	H/325	4.4×10^{-17}	10^{-8}	[80]
Saxitoxin	OPA	CZE	100	H/325	$< 10^{-19}$	$<3 \times 10^{-11}$	[81]
DNA-fragments	FITC	CGE	50 f)	A/488	10^{-20}	-	[82]
Oligonucleotides	JOE	CGE	75	A/-	$< 10^{-19}$	-	[83]
9-Methyl anthracene	-	CEC	10	A/350	7×10^{-19}	1.4×10^{-9}	[84]
Amino acids	CBQCA	MEKC	50	H/442	$1\text{-}7.10^{-17}$	-	[85]
Amino sugars	CBQCA	CZE	50	H/442	$2.4\ 10^{-16}$	-	[86]
Amino acids	FITC	CZE	63	A/488	$2\text{-}8\ 10^{-20}$	10^{-12}	[87]
Mg^{2+}	HQS	CZE	50	A/330	-	10^{-6}	[88]
Mg^{2+},Ca^{2+},Zn^{2+}	HQS	CZE	50	H/325	$2\text{-}10\ 10^{-14}$	$2\text{-}15\ 10^{-6}$	[88]
Nucleotides	DNS	MEKC	20	A/350	$6\ 10^{-18}$	10^{-9}	[89]
Nucleotides	DNS	MEKC	50	A/350	-	10^{-10}	[89]

a) DNS = Dansyl, CBQCA = 3-(4-Carboxybenzoyl)-2-quinoline-carboxaldehyde, OPA = o-Phthaldialdehyde, FITC = Fluorescein isothiocyanate, JOE = Fluorescein based dye [83], HQS = 8-Hydroxy-quinoline-5-sulfonic acid, NDA = Naphthalene dicarboxaldehyde
b) CZE = Capillary Zone Electrophoresis, MEKC = Micellar Electrokinetic Chromatography, CGE = Capillary Gel Electrophoresis, CEC = Capillary Electrochromatography
c) A = Argon ion, H = He/Cd
d) MDL = Mass Detection Limit
e) CDL = Concentration Detection Limit
f) With sheath flow cuvette

As most analytes are not fluorescing, derivatization is necessary. Very often, charge bearing groups, like the amino function, are involved or the fluorophore itself contains charged groups. This may necessitate adaptation of analytical conditions. With proteins, peptides and other multifunctional molecules, complete derivatization is difficult and the resulting structural multiplicity can complicate analysis. Therefore, some researchers have preferred to perform analysis with the native species, followed by on- or post-column derivatization [73,79,80,90-92]. It is evident that the reactors must be carefully constructed to limit additional zone broadening, requiring coaxial capillary mixing chambers [80,90] and zero-volume column connectors [79].

9.5 Data Handling and Interpretation of Observed Area

Peak area is a function of the linear velocity of the zone. The slower it moves, the more time the recording device (a digital data system or a pen recorder) will spend observing it. Comparison of peak areas at different applied voltages is an exaggerated but illustrative example of this principle (**Figure 9.3a**). Comparable areas (**Figure 9.3b**) are obtained after correction for linear velocity [93] (compare with eq.(3.20)) :

$$\text{Area}_{\text{corr},i} = \text{Area}_i \cdot v_i \qquad (9.3)$$

Corrected areas must be considered, not only for comparing analyses, but also within a single analysis. For similar response factors and concentrations, the later eluting peak will show a larger observed area.

Electrokinetic analysis requires data analysis by procedures that are not available on classical integrators or general-purpose data acquisition systems : these include not only the type of area correction just mentioned but also calculation of mobilities and correction for electroosmosis. The first releases of commercial automated instrumentation were often coupled to existing LC-data systems. Software upgrades supporting those calculations have been promised by many companies.

9.6 Indirect Detection Methods

9.6.1 Principles of Indirect Detection

The vast majority of analytical applications in liquid chromatography is based on UV or fluorescence detection. Such detection very often requires derivatization of sample components to introduce suitable chromophores. When the high sensitivity of these detection methods is not required, the derivatization step can be avoided by the use of a universal detector. In liquid chromatography, the differential refractive

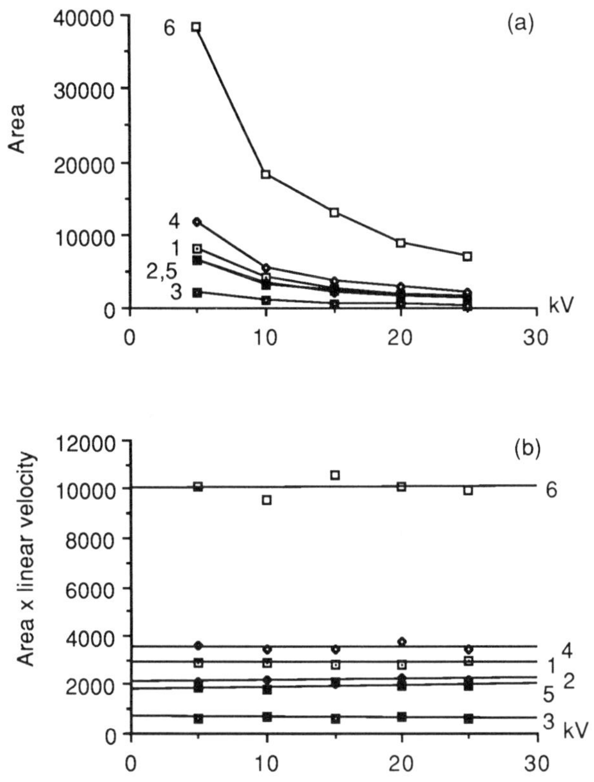

Figure 9.3
Observed (a) and corrected (b) areas in the MEKC analysis of 6 hop acids at different applied voltages. Peaks are numbered according to migration time. Areas are expressed in arbitrary units.

index detector is often used for this purpose. No such simple to use, universal detector is yet available for capillary electrophoresis, even though refractive detectors have been miniaturized and a laser-based version is available for micro LC in 250 μm capillaries [94].

By monitoring laser-beam deflection, sub-millimolar concentrations of native amino acids and sugars have been detected in CZE, using 50 μm capillaries [95].

Such detectors, which are not commercially available, rely on the availability of, and personal acquaintance with, laser technology. Indirect detection [96-98] is an interesting option for universal detection, as almost any existing detector can be used.

To perform indirect detection, a substance is added to the mobile phase that gives a detector response. The solutes do not give the same response as the background (or, preferably, no response at all) and while eluting they will displace part of the active substance which causes a reduction of the signal : negative peaks are observed. Three key factors determine the sensitivity of this detection method : background concentration C_{bg}, transfer ratio TR and dynamic reserve DR.

$$C_{lim} = \frac{C_{bg}}{TR \times DR} \qquad (9.4)$$

The dynamic reserve (DR) is the ratio of background signal to background noise. The stability of the background signal will determine the limiting value of the change in signal that can be detected.

The transfer ratio (TR) is related to the displacement mechanism and expresses the number of active mobile phase molecules that are replaced by one analyte molecule. The simplest displacement mechanism is by volume. It has been remarked that the refractive index detector is in fact an indirect detector, based on volume displacement [97]. A disadvantage of volume displacement is its non-selectivity and consequently low transfer ratio. Of more interest is charge displacement between equally charged background and analyte ions. In this case the transfer ratio is determined by the charge ratio and equally charged analytes will give equal response factors. Consequently, a single calibration curve can be used for all components. Indirect photometric detection was in fact first developed as an alternative to conductivity detection in ion chromatography [96].

9.7.2 Application of Indirect Detection in CZE

In electrophoretic techniques, a background buffer is present and analytes are charged. On using detector-active ions as part of the buffer, charge displacement based indirect detection becomes feasibe. **Table 9.3** lists reported applications.

The conspicuous differences in background concentration are detector-related. The detection limit can be lowered by decreasing the background concentration (see C_{bg} in eq.(9.4)), but this will also reduce the background signal and thus the dynamic range. In fluorescence, the reduction in DR can be compensated by using the increased excitation intensity obtainable from laser light sources [98]. With UV absorbance, however, a sufficiently high background concentration must be used to maintain a stable signal.

Table 9.3 Indirect detection in CZE

Sample	Mode a)	Background b)	CLD.(M) c)	
Inorganic anions, carboxylates	UV	25mM veronal	-	[99]
Amino acids (native)	LIF	0.2 mM salicylate	10^{-5}	[100]
Nucleotides/nucleosides	LIF	0.25 mM salicylate	-	[101]
Inorganic anions	LIF	0.25 mM salicylate	2.10^{-6}	[101]
Proteins (lysozyme)	LIF	0.25 mM salicylate	5.10^{-6}	[101]
Various anions	UV	20 mM benzoate	$<6.10^{-4}$	[102]
Various anions	UV	7 mM sorbate	$<2.10^{-5}$	[102]
Phosphate	LIF	0.25 mM salicylate	2.10^{-7}	[103]
Nucleotide	LIF	1 mM salicylate	-	[103]
Cations	LIF	0.4 mM quinine	$< 10^{-5}$	[104]
Aliphatic amines	LIF	0.4 mM quinine	$<2.5\ 10^{-5}$	[104]
Oligopeptides (tryptic digest)	LIF	0.4 mM quinine	-	[104]
Native amino acids	amp	0.01 mM DHBA d)	2.10^{-8}	[105]
Catecholamines	amp	0.01 mM DHBA	-	[105]
Sugars (native) e)	LIF	1 mM Coumarin 343	-	[106]
Amino acids (native)	LIF	1 mM Coumarin 343	$<7.5\ 10^{-5}$	[106]
Lanthanides	UV	30 mM creatinine	10^{-7}	[107]
Phytate	UV	50 mM benzoate	$< 10^{-5}$	[108]

a) UV = UV absorbance, LIF = Laser-induced fluorescence, amp = Amperometric
b) The concentration is an example of typical working values.
c) CLD = concentration limit of detection, if not determined then the reported concentration is preceded by '<'
d) DHBA = 3,4-Dihydroxybenzylamine
e) At pH 11.5 (pK_a glucose = 12.35)

9.6.3 Possibility of Indirect Detection in MEKC

In the MEKC analysis of neutral compounds, the displacement mechanism is necessarily by volume, which has limited the application of indirect detection. *Yeung* and *Kuhr* have shown the indirect fluorescence detection of aliphatic alcohols against a quinine background [98]. The signal is ascribed to the reduced fluorescence quantum yield when the partitioning of the quinine between both phases was changed by the analyte molecules. The limit of detection is estimated as 10^{-5} M.

In new developments of this kind, sensitivity and universality of detection are often strived for simultaneously, which overlooks those situations where detectability is a problem but sample concentration is not. *Szücs* et al. demonstrated that the sample concentration required for indirect UV detection of neutral compounds in MEKC still allows good separations [109] (**Figure 9.4**).

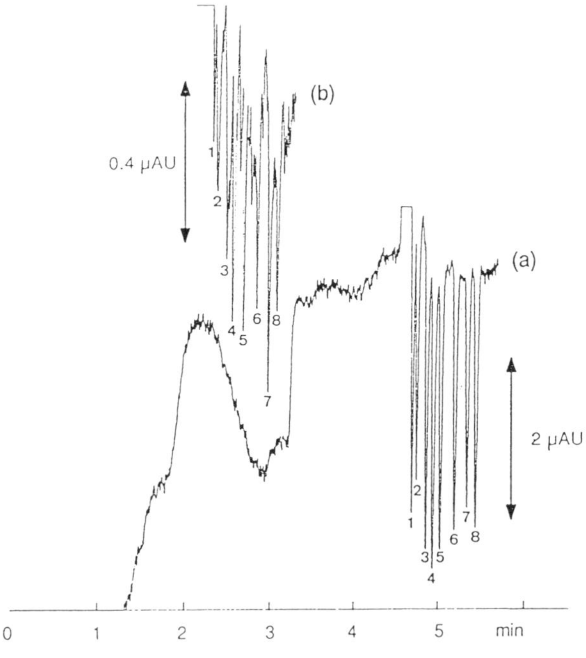

Figure 9.4
MEKC analysis of aliphatic alcohols with indirect UV detection. Capillary : 50 μm x 50/57 cm, deactivated with TMCS. Buffer : 12.5 mM borax with sodium dodecylbenzenesulfonate (SDBS). Applied voltage 15 kV. Detection at 214 nm.
(a) 50 mM SDBS, sample : 1% (v/v) for each alcohol
(b) 4 mM SDBS + 46 mM SDS, sample 0.05% (v/v) for each alcohol
1 = Methanol, 2 = Ethanol, 3 = 2-Propanol, 4 = 1-Propanol, 5 = t-Butanol, 6 = 2-Butanol, 7 = 2-Methyl-1-propanol, 8 = 1-Butanol. *(Reprinted with permission [109] , © Hüthig Verlag)*

9.7 Micro-Preparative Methods

In spite of the miniaturization and the small amounts of sample involved, the increased analytical separation capability has been exploited for micro-preparative purposes. It is required that electric contact is maintained at the column outlet during the collection of the fractions or, alternatively, that the electroosmotic flow can be used to push the sample through a field-free extension of the column.

Using 200 μm ID, gel-filled capillaries, *Hjertén* and *Zhu* [110] collected fractions of pH-indicators for identification by their UV spectra, and tested protein fractions for β-amylase activity. Also with a 200 μm ID gel-filled column and applying overload conditions, *Cohen* et al. isolated the main component from a 800 ng oligonucleotide sample [111]. In subsequent experiments, the difficulty of collecting very narrow zones was recognized and these authors preferred sample collection at low fields after analysis at high field [112].

Rose and *Jorgenson* have constructed a fraction collector [113] and isolated nucleosides and α-chymotrypsin from a 75 μm ID capillary. The biological activity in the enzyme fraction, ca. 25 ng, was demonstrated.

Huang and *Zare* drilled a hole in a 75 μm ID column, in which a frit was made with solder glass. This frit completes the electric circuit prior to the outlet of the column [114], while fractions are collected at the end of the column. They collected ADP fractions with a dilution factor 1:130. By positioning the column outlet against a moving surface [115] they were able to deposit picomole spots of dansylated amino acids on filter paper which could be visualized by their fluorescence.

Instead of a frit, *Fujimoto* et al. [116] used a 'tee'-piece to connect the column end to ground, while collecting fractions from the other outlet and claim to be able to collect fractions of dansylated amino acids from 4 parallel 100 μm ID capillaries simultaneously.

References

[1] D.Kaniansky, P.Havaši, Trends Anal.Chem., **2** (1983) 197.
Instrumentation for capillary isotachophoresis.

[2] P.Gebauer, M.Deml, P.Boček, J.Janák, J.Chromatogr., 267 (1983) 455.
Determination of nitrate,chloride and sulfate in drinking water by capillary free-zone electrophoresis.

[3] T.Tsuda, J.High Resolut.Chromatogr. Chromatogr.Commun., **10** (1987) 622.
Modification of electroosmotic flow with cetyltrimethylammonium bromide in capillary zone electrophoresis.

[4] F.E.P.Mikkers, F.M.Everaerts, Th.P.E.M.Verheggen, J.Chromatogr., **169** (1979) 11.
High-performance zone electrophoresis.

[5] F.Foret, M.Deml, V.Kahle, P.Boček, Electrophoresis (Weinheim), **7** (1986) 430.
On-line fiber optic UV detection cell and conductivity cell for capillary zone electrophoresis.

[6] X.Huang, T.K.J.Pang, M.J.Gordon, R.N.Zare, Anal.Chem., **59** (1987) 2747.
On-column conductivity detector for capillary zone electrophoresis.

[7] X.Huang, M.J.Gordon, R.N.Zare, J.Chromatogr., **425** (1988) 385.
Quantitation of Li^+ in serum by capillary zone electrophoresis with an on-column conductivity detector.

[8] J.L.Beckers, Th.P.E.M.Verheggen, F.M.Everaerts, J.Chromatogr., **452** (1988) 591.
Use of a double-detector system for the measurement of mobilities in zone electrophoresis.

[9] X.Huang, J.A.Luckey, M.J.Gordon, R.N.Zare, Anal.Chem., **61** (1989) 766.
Quantitative determination of low molecular weight carboxylic acids by capillary zone electrophoresis/conductivity detection.

[10] X.Huang, M.J.Gordon, R.N.Zare, J.Chromatogr., **480** (1989) 285.
Effect of electrolyte and sample concentration on the relationship between sensitivity and resolution in capillary zone electrophoresis using conductivity detection.

[11] X.Huang, R.N.Zare, S.Sloss, A.G.Ewing, Anal.Chem., **63** (1991) 189.
End-column detection for capillary electrophoresis.

[12] R.T.Kennedy, M.D.Oates, B.R.Cooper, B.Nickerson, J.W.Jorgenson, Science, **246** (1989) 57.
Microcolumn separations and the analysis of single cells.

[13] A.G.Ewing, T.M.Olefirowicz, Anal.Chem., **62** (1990) 1872.
Capillary electrophoresis in 2 and 5 µm diameter capillaries : application to cytoplasmatic analysis.

[14] R.A.Wallingford, A.G.Ewing, Anal.Chem., 59 (1987) 1762.
Capillary zone electrophoresis with electrochemical detection.

[15] R.A.Wallingford, A.G.Ewing, Anal.Chem., **60** (1988) 258.
Amperometric detection of catechols in capillary zone electrophoresis with normal and micellar solutions.

[16] R.A.Wallingford, A.G.Ewing, Anal.Chem., **60** (1988) 1972.
Capillary zone electrophoresis with electrochemical detection in 12.7 µm diameter columns.

[17] R.A.Wallingford, A.G.Ewing, J.Chromatogr., **441** (1988) 299.
Retention of ionic and non-ionic catechols in capillary zone electrophoresis with micellar solutions.

[18] R.A.Wallingford, A.G.Ewing, Anal.Chem., **61** (1989) 98.
Separation of serotonin from catechols by capillary zone electrophoresis with electrochemical detection.

[19] R.A.Wallingford, P.D.Curry, A.G.Ewing, J.Microcol.Sep., **1** (1989) 23.
Retention of catechols in capillary electrophoresis with micellar and mixed micellar solutions.

[20] C.E.Engstrom-Silverman, A.G.Ewing, J.Microcol.Sep., **3** (1991) 141.
Copper wire amperometric detector for capillary electrophoresis.

[21] D.Scheidegger, S.Röösli, S.Müller, C.Haber, T.Tsuda, W.Simon, Proceedings of the 12th Int.Symp. on Capillary Chromatography, Kobe, Japan (1990) 280.
The potentiometric microelectrode as a detector in fast LC and CZE.

[22] C.Haber, I.Silvestri, S.Röösli, W.Simon, Proceedings of the 13th Int.Symp. on Capillary Chromatography, Riva, Italy (1991) 1701.
Potentiometric detector for capillary zone electrophoresis.

[23] E.D.Lee, W.Mück, J.D.Henion, T.R.Covey, J.Chromatogr., **458** (1988) 313.
On-line capillary zone electrophoresis-ion spray tandem mass spectrometry for the determination of dynorphins.

[24] W.M.Mück, J.D.Henion, J.Chromatogr., **495** (1989) 41.
Determination of leucine enkephalin and methionine enkephalin in equine cerebrospinal fluid by microbore high-performance liquid chromatography and capillary zone electrophoresis coupled to tandem mass spectrometry.

[25] R.D.Smith, H.R.Udseth, J.A.Loo, B.W.Wright, G.A.Ross, Talanta, **36** (1989) 161.
Sample introduction and separation in capillary electrophoresis, and combination with mass spectrometric detection.

[26] J.A.Olivares, N.T.Nguyen, C.R.Yonker, R.D.Smith, Anal.Chem., **59** (1987) 1230.
On-line mass spectrometric detection for capillary zone electrophoresis.

[27] R.D.Smith, J.A.Olivares, N.T.Nguyen, H.R.Udseth, Anal.Chem., **60** (1988) 436.
Capillary zone electrophoresis-mass spectrometry using an electrospray ionization interface.

[28] R.D.Smith, C.J.Barinaga, H.R.Udseth, Anal.Chem., **60** (1988) 1948.
Improved electrospray ionization interface for capillary zone electrophoresis-mass spectrometry.

[29] J.A.Loo, H.K.Jones, H.R.Udseth, R.D.Smith, J.Microcol.Sep., **1** (1989) 223.
Capillary zone electrophoresis-mass spectrometry with electrospray ionization of peptides and proteins.

[30] J.A.Loo, H.R.Udseth, R.D.Smith, Anal.Biochem., **179** (1989) 404.
Peptide and protein analysis by electrospray ionization-mass spectrometry and capillary electrophoresis-mass spectrometry.

[31] R.D.Smith, J.A.Loo, C.J.Barinaga, C.G.Edmonds, H.R.Udseth, J.Chromatogr., **480** (1989) 211.
Capillary zone electrophoresis and isotachophoresis-mass spectrometry of polypeptides and proteins based upon an electropspray ionization interface.

[32] C.G.Edmonds, J.A.Loo, C.J.Barinaga, H.R.Udseth, R.D.Smith, J.Chromatogr., **474** (1989) 21.
Capillary electrophoresis-electrospray ionization-mass spectrometry.

[33] R.D.Smith, J.A.Loo, C.G.Edmonds, C.J.Barinaga, H.R.Udseth, J.Chromatogr., **516** (1990) 157.
Sensitivity considerations for large molecule detection by capillary electrophoresis-electrospray ionization mass spectrometry.

[34] R.D.Smith, J.A.Loo, C.G.Edmonds, C.J.Barinaga, H.R.Udseth, Anal.Chem., **62** (1990) 882.
New developments in biochemical mass spectrometry : Electrospray ionization.

[35] R.D.Smith, S.M.Fields, J.A.Loo, C.J.Barinaga, H.R.Udseth, C.G.Edmonts, Electrophoresis (Weinheim), **11** (1990) 709.
Capillary isotachophoresis with UV and tandem mass spectrometric detection for peptides and proteins.

[36] M.A.Moseley, L.J.Deterding, K.B.Tomer, J.W.Jorgenson, Anal.Chem., **63** (1991) 109.
Determination of bioactive peptides using capillary zone electrophoresis/mass spectrometry.

[37] R.M.Caprioli, W.T.Moore, M.Martin, B.B.DaGue, K.Wilson, S.Moring, J.Chromatogr., **480** (1989) 247.
Coupling capillary zone electrophoresis and continuous-flow fast atom bombardment mass spectrometry for the analysis of peptide mixtures.

[38] R.M.Caprioli, Anal.Chem., **62** (1990) 477A.
Continuous-flow fast atom bombardment mass spectrometry.

[39] M.A.Moseley, L.J.Deterding, K.B.Tomer, J.W.Jorgenson, J.Chromatogr., **480** (1989) 197.
Coupling of capillary zone electrophoresis and capillary liquid chromatographic with coaxial continuous flow fast bombardment tandem sector mass spectrometry.

[40] M.A.Moseley, L.J.Deterding, K.B.Tomer, J.W.Jorgenson, J.Chromatogr., **516** (1990) 167.
Capillary zone electrophoresis-mass spectrometry using a coaxial continuous-flow fast atom bombardment interface.

[41] N.J.Reinhoud, E.Schröder, U.R.Tjaden, W.M.A.Niessen, M.C. Ten Noever de Brauw, J. Van der Greef, J.Chromatogr., **516** (1990) 147.
Static and scanning array detection in capillary electrophoresis-mass spectrometry.

[42] S.L.Pentoney, R.N.Zare, J.F.Quint, J.Chromatogr., **480** (1989) 259.
Semiconductor radioisotope detector for capillary electrophoresis.

[43] S.L.Pentoney, R.N.Zare, J.F.Quint, Anal.Chem., **61** (1989) 1642.
On-line radioisotope detection for capillary electrophoresis.

[44] K.D.Altria, C.F.Simpson, A.K.Bharij, A.E.Theobald, Electrophoresis (Weinheim), **11** (1990) 732.
A gamma-ray detector for capillary zone electrophoresis and its use in the analysis of some radiopharmaceuticals.

[45] R.W.Hallen, C.B.Shumate, W.F.Siems, T.Tsuda, H.H.Hill, J.Chromatogr., **480** (1989) 233.
Preliminary investigation of ion mobility spectrometry after capillary electrophoretic introduction.

[46] F.J.Yang, J.High Resolut.Chromatogr. Chromatogr.Commun., **4** (1981) 83.
On-column detection using a fused silica column.

[47] J.W.Jorgenson, K.D.Lukacs, J.High Resolut.Chromatogr.Chromatogr.Commun., **4** (1981) 230. Zone electrophoresis in open-tubular glass capillaries : Preliminary data on performance.
[48] X.Huang, W.F.Coleman, R.N.Zare, J.Chromatogr., **480** (1989) 95. Analysis of factors causing peak broadening in capillary zone electrophoresis.
[49] K.Otsuka, S.Terabe, J.Chromatogr., **480** (1989) 91. Extra-column effects in high-performance capillary electrophoresis.
[50] H.K.Jones, N.T.Nguyen, R.D.Smith, J.Chromatogr., **504** (1990) 1. Variance contributions to band spread in capillary zone electrophoresis.
[51] T.Wang, R.A.Hartwick, P.B.Champlin, J.Chromatogr., **462** (1989) 147. Adjustable aperture-width detector cell for on-column detection in capillary zone electrophoresis.
[52] J.A.Lux, U.Häusig, G.Schomburg, J.High Resolut.Chromatogr., **13** (1990) 373. Generation of windows in fused silica capillaries for in-column detection of UV-absorption or fluorescence in capillary electrophoresis or HPLC.
[53] R.M.McCormick, R.J.Zagursky, Anal.Chem., **63** (1991) 750. Polyimide stripping device for producing detection windows on fused-silica tubing used in capillary electrophoresis.
[54] Y.Walbroehl, J.W.Jorgenson, J.Chromatogr., **315** (1984) 135. On-column UV absorption detector for open tubular capillary zone electrophoresis.
[55] J.S.Green, J.W.Jorgenson, J.Liq.Chromatogr, **12** (1989) 2527. Design of a variable wavelength UV absorption detector for on-column detection in capillary electrophoresis and comparison to a fixed wavelength UV absorption detector.
[56] J.Vindevogel, G.Schuddinck, C.Dewaele, M.Verzele, J.High Resolut. Chromatogr. Chromatogr.Commun., **11** (1988) 317. Simple instrument modification for packed fused silica capillary or micro-LC : Evaluation of the use of tubular UV-detection cells.
[57] A.E.Bruno, E.Gassmann, N.Pericles, K.Anton, Anal.Chem., **61** (1989) 876. On-column capillary flow cell utilizing optical waveguides for chromatographic applications.
[58] R.J.Nelson, A.Paulus, A.Guttman, A.S.Cohen, B.L.Karger, J.Chromatogr., **480** (1989) 111. Use of Peltier thermoelectric devices to control column temperature in high-performance capillary electrophoresis.
[59] S.E.Moring, J.C.Colburn, P.D.Grossman, H.H.Lauer, LC*GC, **8**(1) (1990) 34 (also : LC*GC Int., **3**(2) (1990) 46). Analytical aspects of an automated capillary electrophoresis system.
[60] T.Tsuda, Y.Kobayashi, J.Chromatogr., **515** (1990) 357. Use of a focusing cylindrical lens for increasing sensitivity in the optical detector of a capillary flow-through cell.
[61] T.Tsuda, Y.Kobayashi, A.Hori, T.Matsumoto, O.Suzuki, J.Microcol.Sep., **2** (1990) 21. Separation of polyamines in rat tissues by capillary electrophoresis.
[62] T.Tsuda, J.V.Sweedler, R.N.Zare, Anal.Chem., **62** (1990) 2149. Rectangular capillaries for capillary zone electrophoresis.
[63] J.P.Chervet, M.Ursem, J.P.Salzmann, R.W.Vannoort, J.High Resolut.Chromatogr.Chromatogr. Commun., **12** (1989) 278. Ultra-sensitive UV detection in micro separations.
[64] J.P.Chervet, R.E.J.Van Soest, M.Ursem, J.Chromatogr., **543** (1991) 439. Z-shaped flow cell for UV detection in capillary electrophoresis.
[65] I.H.Grant, W.Steuer, J.Microcol.Sep., **2** (1990) 74. Extended path length UV absorbance detector for capillary zone electrophoresis.
[66] M.J.Sepaniak, D.F.Swaile, A.C.Powell, J.Chromatogr., **480** (1989) 185. Instrumental developments in micellar electrokinetic capillary chromatography.
[67] S.Kobayashi, T.Ueda, M.Kikumoto, J.Chromatogr., **480** (1989) 179. Photodiode array detection in high-performance capillary electrophoresis.
[68] J.Vindevogel, P.Sandra, L.C.Verhagen, J.High Resolut.Chromatogr., **13** (1990) 295. Separation of hop bitter acids by capillary zone electrophoresis and micellar electrokinetic chromatography with UV-diode array detection.
[69] J.Vindevogel, R.Szücs, P.Sandra, L.C.Verhagen, J.High Resolut.Chromatogr., **14** (1991) 584. Analysis of beer iso-α-acids by micellar electrokinetic chromatography and multi-wavelength UV detection.

[70] J.S.Green, J.W.Jorgenson, J.Chromatogr., **352** (1986) 337.
Variable-wavelength on-column fluorescence detector for open-tubular zone electrophoresis.
[71] G.J.M.Bruin, J.P.Chang, R.H.Kuhlman, K.Zegers, J.C.Kraak, H.Poppe, J.Chromatogr., **471** (1989) 429.
Capillary zone electrophoretic separations of proteins in polyethylene glycol-modified capillaries.
[72] Y.Kurosu, T.Sasaki, M.Saito, J.High Resolut.Chromatogr., **14** (1991) 186.
Fluorescence detection with immersed flow cell in capillary electrophoresis.
[73] M.Albin, R.Weinberger, E.Sapp, S.Moring, Anal.Chem., **63** (1991) 417.
Fluorescence detection in capillary electrophoresis : evaluation of derivatizing reagents and techniques.
[74] S.Wu, N.J.Dovichi, J.Chromatogr., **480** (1989) 141.
High-sensitivity fluorescence detector for fluorescein isothiocyanate derivatives of amino acids separated by capillary zone electrophoresis.
[75] P.Gozel, E.Gassmann, H.Michelsen, R.N.Zare, Anal.Chem., **59** (1987) 44.
Electrokinetic resolution of amino acid enantiomers with copper(II)-aspartame support electrolyte.
[76] M.C.Roach, P.Gozel, R.N.Zare, J.Chromatogr., **442** (1988) 129.
Determination of methotrexate and its major metabolite, 7-hydroxymethtrexate, using capillary zone electrophoresis and laser-induced fluorescence detection.
[77] Y.F.Cheng, N.J.Dovichi, Science, **242** (1988) 562.
Subattomole amino acid analysis by capillary zone electrophoresis and laser-induced fluorescence.
[78] B.Nickerson, J.W.Jorgenson, J.High Resolut.Chromatogr. Chromatogr. Commun., **11** (1988) 533.
High speed capillary zone electrophoresis with laser induced fluorescence detection.
[79] S.L.Pentoney, X.Huang, D.S.Burgi, R.N.Zare, Anal.Chem., **60** (1988) 2625.
On-line connector for microcolumns : Application to the on-column o-phthaldialdehyde derivatization of amino acids separated by capillary zone electrophoresis.
[80] B.Nickerson, J.W.Jorgenson, J.Chromatogr., **480** (1989) 157.
Characterization of a post-column reaction-laser-induced fluorescence detector for capillary zone electrophoresis.
[81] B.W.Wright, G.A.Ross, R.D.Smith, J.Microcol.Sep., **1** (1989) 85.
Capillary zone electrophoresis with laser fluorescence detection of marine toxins.
[82] H.Swerdlow, S.Wu, H.Harke, N.J.Dovichi, J.Chromatogr., **516** (1990) 61.
Capillary gel electrophoresis for DNA sequencing. Laser-induced fluorescence detection with the sheath-flow cuvette.
[83] A.S.Cohen, D.R.Najarian, B.L.Karger, J.Chromatogr., **516** (1990) 49.
Separation and analysis of DNA sequence reaction products by capillary gel electrophoresis.
[84] W.D.Pfeffer, E.S.Yeung, Anal.Chem., **62** (1990) 2178.
Open-tubular liquid chromatography with surfactant enhanced electroosmotic flow.
[85] J.Liu, Y.-Z.Hsieh, D.Wiesler, M.Novotny, Anal.Chem., **63** (1991) 408.
Design of 3-(4-carboxybenzoyl)-2-quinolinecarboxaldehyde as a reagent for ultrasensitive determination of primary amines by capillary electrophoresis using laser fluorescence detection.
[86] J.Liu, O.Shirota, M.Novotny, Anal.Chem., **63** (1991) 413.
Capillary electrophoresis of amino sugars with laser-induced fluorescence detection.
[87] J.V.Sweedler, J.B.Shear, H.A.Fishman, R.N.Zare, R.H.Scheller, Anal.Chem., **63** (1991) 496.
Fluorescence detection in capillary zone electrophoresis using a charge-coupled device with time-delayed integration.
[88] D.F.Swaile, M.J.Sepaniak, Anal.Chem., **63** (1991) 179.
Determination of metal ions by capillary zone electrophoresis with on-column chelation using 8-hydroxy-quinoline-5-sulfonic acid.
[89] T.Lee, E.S.Yeung, M.Sharma, J.Chromatogr., **565** (1991) 197.
Micellar electrokinetic capillary chromatographic separation and laser-induced fluorescence detection of 2'-deoxynucleoside 5'-monophosphates of normal and modified bases.
[90] D.J.Rose, J.W.Jorgenson, J.Chromatogr., **447** (1988) 117.
Post-capillary fluorescence detection in capillary zone electrophoresis using o-phthaldialdehyde.
[91] T.Tsuda, Y.Kobayashi, A.Hori, T.Matsumoto, O.Suzuki, J.Chromatogr., **456** (1988) 375.
Post-column detection for capillary zone electrophoresis.

[92] D.J.Rose, J.Chromatogr., **540** (1991) 343.
Free-solution reactor for post-column fluorescence detection in capillary zone electrophoresis.
[93] W.J.Lambert, D.L.Middleton, Anal.Chem., **62** (1990) 1585.
pH hysteresis effect with silica capillaries in capillary zone electrophoresis.
[94] D.J.Bornhop, T.G.Nolan, N.J.Dovichi, J.Chromatogr., **384** (1987) 181.
Subnanoliter laser-based refractive index detector for 0.25 mm I.D. microbore liquid chromatography. Reversed-phase separation of nanogram amounts of sugars.
[95] J.Pawliszyn, Anal.Chem., **60** (1988) 2796.
Nanoliter volume sequential differential concentration gradient detector.
[96] H.Small, T.E.Miller, Anal.Chem., **54** (1982) 462.
Indirect photometric chromatography.
[97] E.S.Yeung, Acc.Chem.Res., **22** (1989) 125.
Indirect detection methods : looking for what is not there.
[98] E.S.Yeung, W.G.Kuhr, Anal.Chem., **63** (1991) 275A.
Indirect detection methods for capillary separations.
[99] S.Hjertén, K.Elenbring, F.Kilár, J.-L.Liao, A.J.C.Chen, C.J.Siebert, M.-D.Zhu, J.Chromatogr., **403** (1987) 47.
Carrier-free zone electrophoresis, displacement electrophoresis and isoelectric focusing in a high-performance electrophoresis apparatus.
[100] W.G.Kuhr ,E.S.Yeung, Anal.Chem., **60** (1988) 1832.
Indirect fluorescence detection of native amino acids in capillary zone electrophoresis.
[101] W.G.Kuhr, E.S.Yeung, Anal.Chem., **60** (1988) 2642.
Optimization of sensitivity and separation in capillary zone electrophoresis with indirect fluorescence detection.
[102] F.Foret, S.Fanali, L.Ossicini, P.Boček, J.Chromatogr., **470** (1989) 299.
Indirect photometric detection in capillary zone electrophoresis.
[103] L.Gross, E.S.Yeung, J.Chromatogr., **480** (1989) 169.
Indirect fluorimetric detection and quantification in capillary zone electrophoresis of inorganic anions and nucleotides.
[104] L.Gross, E.S.Yeung, Anal.Chem., **62** (1990) 427.
Indirect fluorimetric detection of cations in capillary zone electrophoresis.
[105] T.M.Olefirowicz, A.G.Ewing, J.Chromatogr., **499** (1990) 713.
Capillary electrophoresis with indirect amperometric detection.
[106] T.W.Garner, E.S.Yeung, J.Chromatogr., **515** (1990) 639.
Indirect fluorescence detection of sugars by capillary zone electrophoresis with visible layer excitation.
[107] F.Foret, S.Fanali, A.Nardi, P.Boček, Electrophoresis (Weinheim), **11** (1990) 780.
Capillary zone electrophoresis of rare earth metals with indirect UV absorbance detection.
[108] A.Nardi, M.Cristalli, C.Desiderio, L.Ossicini, S.K.Shukla, S.Fanali, Proceedings of the 13th Symposium on Capillary Chromatography, Riva del Garda (Italy) (1991) 1181.
Indirect UV photometric detection in capillary zone electrophoresis for the determination of phytate in soybean.
[109] R.Szücs, J.Vindevogel, P.Sandra, J.High Resolut.Chromatogr., **14** (1991) 692.
Micellar electrokinetic chromatography of aliphatic compounds with indirect UV detection.
[110] S.Hjertén, M.-D.Zhu, J.Chromatogr., **327** (1985) 157.
Micropreparative version of high-performance electrophoresis. The electrophoretic counterpart of narrow-bore high-performance liquid chromatography.
[111] A.S.Cohen, A.Paulus, B.L.Karger, Chromatographia, **24** (1987) 15.
High-performance capillary electrophoresis using open tubes and gels.
[112] A.Guttman, A.S.Cohen, D.N.Heiger, B.L.Karger, Anal.Chem., **62** (1990) 137.
Analytical and micropreparative ultrahigh resolution of oligonucleotides by polyacrylamide gel high-performance capillary electrophoresis.
[113] D.J.Rose, J.W.Jorgenson, J.Chromatogr., **438** (1988) 23.
Fraction collector for capillary zone electrophoresis.
[114] X.Huang, R.N.Zare, Anal.Chem., **62** (1990) 443.
Use of an on-column frit in capillary zone electrophoresis : sample collection.

[115] X.Huang, R.N.Zare, J.Chromatogr., **516** (1990) 185.
Continuous sample collection in capillary zone electrophoresis by coupling the outlet of a capillary to a moving surface.
[116] C.Fujimoto, Y.Muramatsu, M.Suzuki, K.Jinno, J.High Resolut.Chromatogr., **14** (1991) 178.
Capillary electrophoresis separation of amino acids : fraction collection.

10 Efficiency in MEKC

10.1 In-Column Contributions to Zone Broadening

Terabe et al. have comprehensively and theoretically evaluated the contributions to zone broadening in MEKC [1]. Much of the discussion presented here is based on their treatment. After experimental confirmation, obtained in 50 µm ID capillaries with neutral test solutes (**Figure 10.1**), they concluded that axial diffusion was more important than resistance to mass transfer and micelle micro-heterogeneity, at least for capacity factors up to 5 and electroosmotic velocities up to 2 mm/sec. Others have observed and discussed non-equilibrium effects [2,3], leading to a Golay-type dependence of plate height on linear velocity or applied voltage.

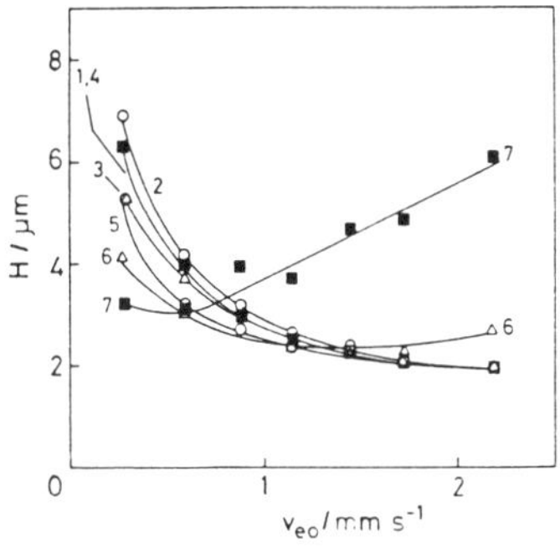

Figure 10.1
Effect of electroosmotic mobility on the plate height in MEKC. Column : 50 µm x 50/65 cm. Buffer : 50 mM SDS in 100 mM phosphate/borate (pH 7.0). The applied field is in the range 100-330 V/cm.
(1) Resorcinol, (2) Phenol, (3) p-Nitroaniline, (4) Nitrobenzene, (5) Toluene, (6) 2-Naphthol, (7) Sudan III. *(Reprinted with permission [1] © American Chemical Society)*

10.1.1 Axial or Longitudinal Diffusion

In conventional liquid chromatography, the diffusion in the stationary phase is considered negligible. In MEKC however, the micellized solute will show the same diffusion characteristics as the micelle [1] :

$$\sigma^2_{diff} = \sigma^2_{diff,aq} + \sigma^2_{diff,MC} = \frac{1}{1+\tilde{k}'} 2\, D_m\, t_R + \frac{\tilde{k}'}{1+\tilde{k}'} 2\, D_{m,MC}\, t_R \qquad (10.1)$$

and, as t_R is also a funtion of $\tilde{k}'$

$$\sigma^2_{diff} = \frac{2\, t_0}{1 + \frac{t_0}{t_{MC}}\tilde{k}'}\left(D_m + \tilde{k}'\, D_{m,MC}\right) \qquad (10.2)$$

and[1]

$$h_{diff} = \frac{\sigma^2_{diff}}{L} = \frac{2\left(D_m + \tilde{k}'\, D_{m,MC}\right)}{1 + \frac{t_0}{t_{MC}}\tilde{k}'}\,\frac{1}{v_{eo}} = \frac{2\left(D_m + \tilde{k}'\, D_{m,MC}\right)}{B'}\,\frac{1}{v_{eo}} = \frac{B}{v_{eo}} \qquad (10.3)$$

In their experiments, *Terabe* et al. [1] found a linear relationship between plate height and the inverse of the electroosmotic velocity, according to eq.(10.3). $D_{m,MC}$ is a function of concentration and added salt [4], for SDS micelles the reported values range from 5×10^{-11} to 10^{-10} m^2/sec [4,5], the lower value being representative for MEKC conditions. For the typical, relatively small molecules, to which MEKC is applicable, D_m is one order of magnitude larger and, at least for low capacity factors, the influence of diffusion in the micellar phase is negligible.

10.1.2 Resistance to Mass Transfer

As long as axial diffusion is predominating, highest efficiency is obtained at high velocity. Resistance to (radial) transport is the reason why efficiency deteriorates at higher velocities. This is expressed in the Golay equation [6] :

$$h = \frac{B}{v} + C\, v \qquad (10.4)$$

In MEKC, the contributions from mass transfer kinetics to plate height are small, explaining the superior efficiency obtained in comparison to liquid chromatography.

---

[1] *Typical values are summarized in **Table 10.1**.*

A first factor is the micellar mass transfer [3] or the effect of sorption-desorption kinetics [1]. Using literature values for the desorption rates k_d, ranging from 4.1 x 10^2 sec^{-1} (perylene) to 4.4 x 10^6 sec^{-1} (benzene) [7], *Terabe* et al. showed that, even under unfavorable conditions, the contribution to plate height would be less than 1.3 μm [1] :

$$h_{MC} = \frac{2\left(1 - \frac{t_0}{t_{MC}}\right)^2 \tilde{k}'}{\left(1 + \frac{t_0}{t_{MC}}\tilde{k}'\right)\left(1 + \tilde{k}'\right)^2} \frac{v_{eo}}{k_d} = C'_1 \frac{v_{eo}}{k_d} = C_1\, v_{eo} \qquad (10.5)$$

The reduced importance of micellar mass transfer can be easily understood. Micelles are small, dynamic entities. The micelle radius is close to the length of the hydrophobic chain [8]; for SDS this is only 2.4 to 2.7 nm [9] [1]. The micelle residence time ranges from a few to 250 μsec, while the micelle itself has a lifetime of 100 to 1000 msec [8]. The risk of irreversible 'trapping' or slow exchange is strongly reduced. Although, according to *Terabe* et al., this mass transfer contribution is already essentially low, some have ascribed observed plate count increases, after the addition of organic modifiers [11] or short-chain cosurfactants [12], to changes in micelle structure.

A second factor is intermicelle diffusion. Its effect on plate height was shown by *Terabe* et al. [1] to be proportional to the square of the intermicelle distance d_{MC} and completely negligible in 50 mM SDS solutions ($d_{MC} \approx 10$ nm) :

$$h_{aq} = \left(\frac{\tilde{k}'}{1 + \tilde{k}'}\right)^2 \frac{\left(1 - \frac{t_0}{t_{MC}}\right)^2}{1 + \frac{t_0}{t_{MC}}\tilde{k}'} \frac{d_{MC}^2\, v_{eo}}{4\, D_m} = C'_2 \frac{d_{MC}^2\, v_{eo}}{4\, D_m} = C_2\, v_{eo} \qquad (10.6)$$

Reported amelioration of efficiency with increasing surfactant concentration [2,13,14] invariably results from comparing extremely low surfactant concentrations (15 mM or less, close to the CMC) with more normal levels (50-100 mM), and must be attributed to causes other than the effect of the decreased d_{MC} [2].

---

[1] *For comparison, thinnest films in gas chromatography are still approximately 100 nm [10].*

[2] *It can be calculated that d_{MC} is only doubled or tripled (depending on the actual CMC) on reducing the surfactant from 100 to 10 mM.*

A third factor of resistance to lateral transport is related to the radial distribution of the analytes in the column. The contribution to plate height due to the electroosmotic flow profile has already been discussed in relation to electrochromatography and can be considered negligible provided that $k'_{wall} \approx 0$ (not to be confused with $\tilde{k}'$). If this is not true, as has been observed when partition occurs with a TMCS-coated column wall [11], the relatively large column diameter causes a sharp drop in efficiency. Another possibility to obtain radial differences in velocity is by the influence of the temperature gradient on viscosity, which in turn affects the electrophoretic mobility of the micellized solutes. Using a temperature gradient description by *Hjertén* [15], *Terabe* et al. calculated that the contribution to plate height is again negligible in 50 μm ID columns [1] [1] :

$$h_{\Delta T} = \frac{\left(1 - \frac{t_0}{t_{MC}}\right)\tilde{k}'}{24\left(D_m + \tilde{k}' D_{m,MC}\right)} \frac{E^4 \kappa^2 r_c^6 B_\eta^2}{64 \lambda_T^2 T_c^4} v_{eo} = \frac{C'_3}{C''_3} C'''_3 v_{eo} = C_3 v_{eo} \qquad (10.7)$$

Table 10.1 Typical values for several plate height contributions (for $t_{MC}/t_0 = 5$ and $\tilde{k}' = \tilde{k}'_{opt} = 2.24$)

	Eq.	Relevant values			h (μm) at v_{eo} = 1 mm/sec
h_{diff}	10.3	$B' = 0.45$	$D_m = 10^{-3}$ mm²/sec	$D_{m,MC} = 10^{-4}$ mm²/sec	5.4
h_{MC}	10.4	$C'_1 = 0.19$	$k_d = 10^3$ sec^{-1}		0.2
h_{aq}	10.6	$C'_2 = 0.21$	$D_m = 10^{-3}$ mm²/sec	$d_{MC} = 10^{-5}$ mm	5×10^{-6}
$h_{\Delta T}$	10.7	$C'_3 = 1.79$	$C''_3 = 0.03$ mm²/sec a)		< 1.0 b)
h_{disp}	10.8	$C'_4 = 0.00003$	$k_d = 10^3$ sec^{-1}		$3.x \times 10^{-5}$
h_{inj}	8.1	$l_{inj} = 1$ mm	$L_d = 500$ mm		0.17

a) Same diffusion coefficient values as for h_{diff}
b) See **Figure 10.2**

[1] *Using $I = E \pi r^2 \kappa$, the original equation has been rewritten to avoid interrelated quantities like current and column radius.*

As *Terabe* et al. did not consider the effect of the column diameter, nor the fact that increasing linear velocity through applied voltage introduces a fifth-power dependence on the field strength, the effects of column diameter and applied voltage have been recalculated (**Figure 10.2**). In no case can a significant contribution to plate height be found. According to eq.(10.7), thermal gradient effects cannot account for the increase of plate height with applied voltage that has been observed by others in 60 and 75 μm ID columns [2,16].

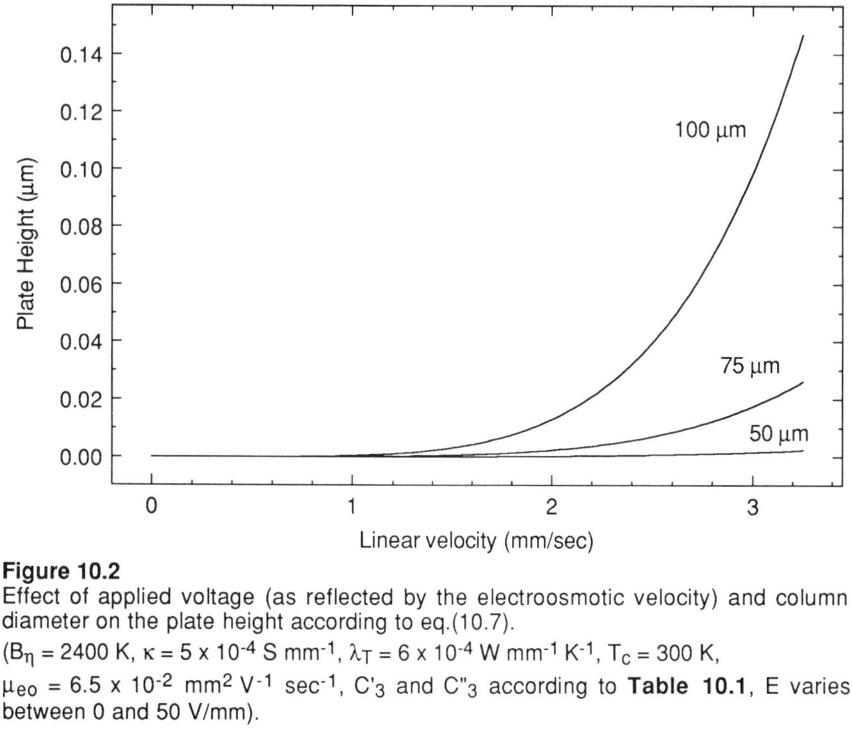

Figure 10.2
Effect of applied voltage (as reflected by the electroosmotic velocity) and column diameter on the plate height according to eq.(10.7).
(B_η = 2400 K, $\kappa = 5 \times 10^{-4}$ S mm^{-1}, $\lambda_T = 6 \times 10^{-4}$ W mm^{-1} K^{-1}, T_c = 300 K, $\mu_{eo} = 6.5 \times 10^{-2}$ mm^2 V^{-1} sec^{-1}, C'_3 and C''_3 according to **Table 10.1**, E varies between 0 and 50 V/mm).

10.1.3 Micelle Polydispersity or Microheterogeneity

Assuming an aggregation number of 64 for SDS micelles, with a corresponding standard deviation of 13 [8], *Terabe* et al. [1] have calculated that the relative standard deviation of the micelle electrophoretic mobility is less than 0.5%[1]. Using random-walk theory, considering that the micelle residence time of the solute is shorter than the micelle lifetime, this leads to a negligible plate height contribution for low capacity factors:

$$h_{disp} = \frac{(0.0053)^2 \left(1 - \frac{t_0}{t_{MC}}\right)^2 \tilde{k}'}{\left(1 + \frac{t_0}{t_{MC}} \tilde{k}'\right)} \frac{v_{eo}}{k_d} = C'_4 \frac{v_{eo}}{k_d} = C_4 v_{eo} \tag{10.8}$$

For this type of process, a pseudo-diffusion contribution would be expected, and indeed the mass transfer-like dependence on v_{eo} is a mathematical artefact, resulting from the following substitution, aimed at introducing a relative standard deviation for $v_{ep,MC}$ (see eq.(38) in [1]) :

$$\frac{\sigma^2_{ep,MC}}{v_{eo}} = \frac{\sigma^2_{ep,MC}}{v_{eo}} \cdot \frac{v^2_{eo}}{v^2_{ep,MC}} \left(1 - \frac{t_0}{t_{MC}}\right)^2 = \left(\frac{\sigma_{ep,MC}}{v_{ep,MC}}\right)^2 v_{eo} \left(1 - \frac{t_0}{t_{MC}}\right)^2 \tag{10.9}$$

With very high capacity factors, e.g. for Sudan III, microheterogeneity was suspected to be a cause of band broadening, resulting in a moderate increase of h with linear velocity or applied voltage. However, experimental values showed a much stronger dependence than predicted. Using back-and-forth electrophoresis between twin detectors [17], it was later demonstrated that more than half of the observed zone broadening was due to heterogeneity of the tracer (caused by impurities) instead of the micelles.

10.2 Thermal Contributions to Zone Broadening

Terabe et al. have warned that their calculations on plate height contributions predict plate heights less than 5-6 μm [1]. In the case of larger values, other plate

---

[1] *Terabe et al. reported 2.6% which is a calculation error (σ_n/n was not considered in eq.(43) from [1]). The factor 0.026 must be replaced by 0.0053 in eq.(44) in [1] and, moreover, must be squared (according to eq.(38) in [1]). Equation (10.8) has been adapted accordingly. This further reduces the significance of h_{disp} and its contribution at large $\tilde{k}'$ values (Figure 6c from [1]), leaving the relatively strong C term for Sudan III unexplained.*

height contributions must be suspected. The effects of detection and especially, injection length have already been discussed in a general way and apply equally well to MEKC. An example of volume overloading has already been shown in **Figure 8.4** and **8.5**.

Nevertheless, others have reported non-equilibrium effects in the dependence of plate height on applied voltage (**Figure 10.3**). In these cases where C values were not negligible [2,16], they were more than two orders of magnitude higher than predicted by the theoretical considerations of *Terabe* et al. [3]. To explain this discrepancy, *Davis* has postulated another form of transchannel mass transfer resistance, related to the variation of the distribution coefficient and capacity factor with the radial temperature gradient [3] (**Figure 10.4**) which is unique for MEKC. But in his *ad hoc* statistical treatment, based on random-walk theory, Δv_R is not explicitly related to radial changes in $\tilde{k}'$ or ΔT. The resulting mass transfer term predicts the non-equilibrium that is observed in some cases but fails to explain why it is absent in other cases.

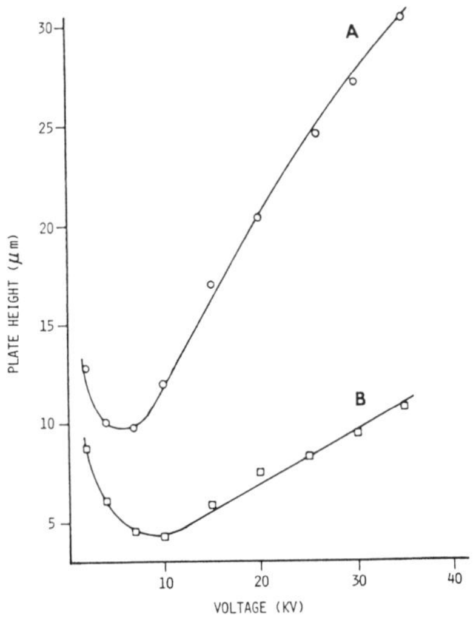

Figure 10.3
Plate height vs. applied voltage in the MEKC analysis of NBD-cyclohexylamine (A) and NBD-ethylamine (B). Column : 75 μm x 75 cm. Buffer : 10 mM SDS in 3 mM phosphate. *(Reprinted with permission [2] © American Chemical Society)*

The discussion on the existence of non-equilibrium conditions in MEKC is based on a remarkably small collection of experimental data, obtained with different compounds and under different conditions. In at least one case, although involved with the MEKC analysis of ions, both the equilibrium and non-equilibrium have been observed (**Figure 10.5**) under comparable conditions [18]. The only difference is the buffer, which is TRIS in the non-equilibrium case and is known to cause selectivity changes for hop acids as a result of in-column heating [18], supporting the zone broadening mechanism proposed by *Davis*.

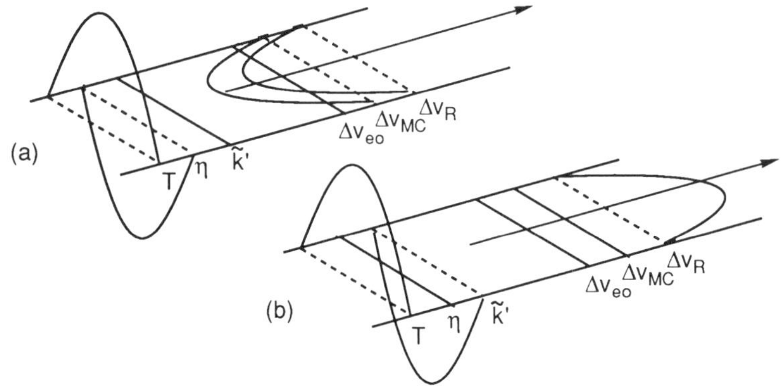

Figure 10.4
Velocity profiles resulting from the effect of the temperature gradient on the viscosity (a) or the capacity factor (b). Trends rather than exact relationships are shown. Case (a) is the conventional way to represent thermal effects in CZE and MEKC. Case (b) is unique for MEKC and has been postulated by *Davis* [3].

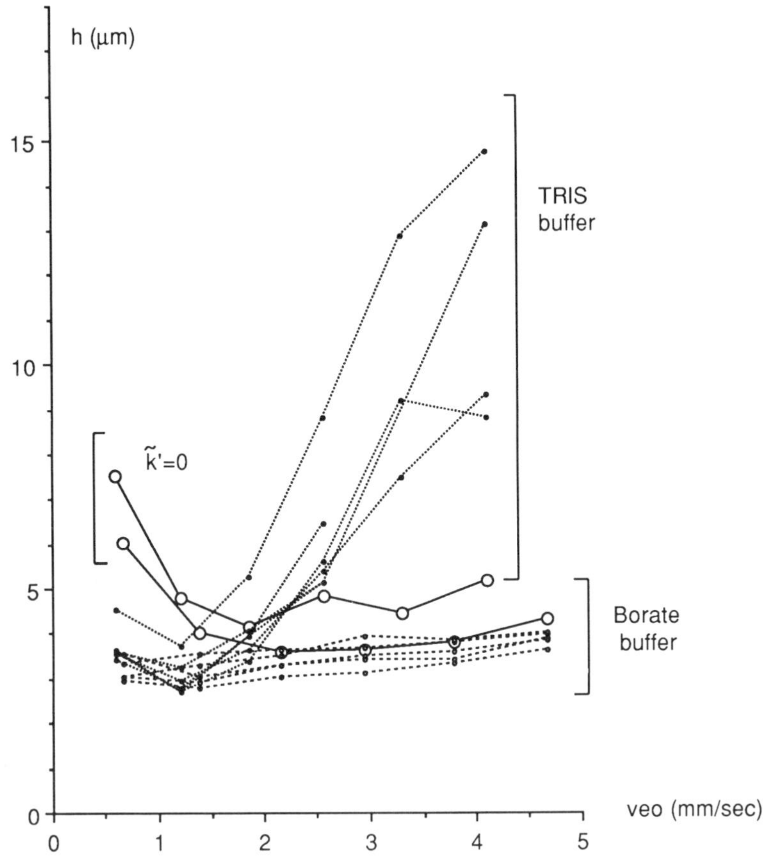

Fig. 10.5
Plate height vs. electroosmotic velocity for hop bitter acids in a borate and a TRIS buffer. Column : 75 µm x 43/50 cm. Buffer : 30 mM SDS in 25 mM borate (pH 8.55) or TRIS (pH 8.40). The applied field is varied between 100 and 600 V/cm. The graph is based on data from [18].

References

[1] S.Terabe, K.Otsuka, T.Ando, Anal.Chem., **61** (1989) 251.
Band broadening in electrokinetic chromatography with micellar solutions and open-tubular capillaries.
[2] M.J.Sepaniak, R.O.Cole, Anal.Chem., **59** (1987) 472.
Column efficiency in micellar electrokinetic capillary chromatography.
[3] J.M.Davis, Anal.Chem., **61** (1989) 2455.
Random-walk theory of nonequilibrium plate height in micellar electrokinetic chromatography.
[4] D.Stigter, R.J.Williams, K.J.Mysels, J.Phys.Chem., **59** (1955) 331.
Micellar self diffusion of sodium lauryl sulfate.
[5] R.J.McGreevy, R.S.Schechter, J.Colloid Interface Sci., **127** (1989) 209.
Influence of n-butanol on the size of sodium dodecyl sulfate micelles.
[6] J.C.Giddings, "Dynamics of Chromatography. Part I : Principles and Theory", Dekker Inc, New York, 1965.
[7] M.Almgren, F.Griesen, J.K.Thomas, J.Am.Chem.Soc., **101** (1979) 279.
Quasi-elastic light scattering study of intermicellar interactions in aqueous sodium dodecyl sulfate solutions.
[8] E.A.G.Aniansson, S.N.Wall, M.Almgren, H.Hoffmann, I.Kielmann, W.Ulbricht, R.Zana, J.Lang, C.Tondre, J.Phys.Chem., **80** (1976) 905.
Theory of the kinetics of micelle equilibria and quantitative interpretation of chemical relaxation studies of micellar solutions of ionic surfactants.
[9] D.E.Dunstan, L.R.White, J.Colloid Interface Sci., **134** (1990) 147.
An electrokinetic study of micellar solutions.
[10] P.Sandra, J.High Resolut.Chromatogr.Chromatogr.Commun., **12** (1989) 273.
Resolution - Column efficiency.
[11] A.T.Balchunas, M.J.Sepaniak, Anal.Chem., **59** (1987) 1466.
Extension of elution range in micellar electrokinetic capillary chromatography.
[12] R.A.Wallingford, P.D.Curry, A.G.Ewing, J.Microcol.Sep., **1** (1989) 23.
Retention of catechols in capillary electrophoresis with micellar and mixed micellar solutions.
[13] A.T.Balchunas, M.J.Sepaniak, Anal.Chem., **60** (1988) 617.
Gradient elution for micellar electrokinetic capillary chromatography.
[14] M.J.Sepaniak, D.F.Swaile, A.C.Powell, R.O.Cole, J.High Resolut.Chromatogr., **13** (1990) 679.
Capillary electrokinetic separations : influence of mobile phase composition on performance.
[15] S.Hjertén, Chromatogr.Rev., **9** (1967) 122.
Free zone electrophoresis.
[16] K.H.Row, W.H.Griest, M.P.Maskarinec, J.Chromatogr., **409** (1987) 193.
Separation of modified nucleic acid constituents by micellar electrokinetic capillary chromatography.
[17] S.Terabe, O.Shibata, T.Isemura, J.High Resolut.Chromatogr., **14** (1991) 52.
Band-broadening evaluation by back-and-forth capillary electrophoresis.
[18] J.Vindevogel, P.Sandra, J.High Resolut.Chromatogr., **14** (1991) 795.
Micellar electrokinetic chromatography of ionic compounds : The analysis of hop bitter acids.

11 Resolution and Resolution Optimization

11.1 Factors Affecting Resolution

Starting with the fundamental resolution equation

$$R_S = \frac{\Delta t_M}{\frac{1}{2}(w_1 + w_2)} = \frac{\Delta t_M}{\overline{w}} \qquad (11.1)$$

and following the same reasoning as in conventional chromatography [1], the resolution between closely eluting peaks ($N \approx N_1 \approx N_2$) is described by [2] :

$$R_S = \frac{\sqrt{N}}{4} \cdot \frac{\alpha - 1}{\alpha} \cdot \frac{\tilde{k}'_2}{1 + \tilde{k}'_2} \cdot \frac{1 - \frac{t_0}{t_{MC}}}{1 + \frac{t_0}{t_{MC}} \cdot \tilde{k}'_1} \qquad (11.2)$$

For a true stationary phase (t_{MC} equal to infinity), eq.(11.2) reduces to the well-known equation in conventional chromatography :

$$R_S = \frac{\sqrt{N}}{4} \cdot \frac{\alpha - 1}{\alpha} \cdot \frac{\tilde{k}'_2}{1 + \tilde{k}'_2} \qquad (11.3)$$

The fourth term in eq.(11.2) is responsible for the particular resolution characteristics in MEKC. Assuming that $\tilde{k}' \approx \tilde{k}'_1 \approx \tilde{k}'_2$ [1], eq.(11.2) can be rewritten as

$$R_S = \frac{\sqrt{N}}{4} \cdot \frac{\alpha - 1}{\alpha} \cdot f_k \qquad (11.4)$$

with

[1] *This assumption is less valid than in conventional chromatography. Near t_{MC}, closely eluting peaks (assumed when deriving eq.(11.2)) can have considerably different $\tilde{k}'$ values. On the other hand, for high $\tilde{k}'$ values, the third term in eq.(11.2) is relatively insensitive to changes in $\tilde{k}'$.*

$$f_k = \frac{\tilde{k}'}{1+\tilde{k}'} \cdot \frac{1 - \dfrac{t_0}{t_{MC}}}{1 + \dfrac{t_0}{t_{MC}} \cdot \tilde{k}'} \qquad (11.5)$$

The contribution of f_k to resolution, as a function of the capacity factor, and for different elution range values, i.e. t_{MC}/t_0 [1], is shown in **Figure 11.1**. A maximum is attained for intermediate capacity factors. Using

$$\frac{\partial f_k}{\partial \tilde{k}'} = 0 \qquad (11.6)$$

it can be shown that this maximum is obtained at

$$\tilde{k}'_{opt} = \sqrt{\frac{t_{MC}}{t_0}} \qquad (11.7)$$

From **Figure 11.1** it could be deduced that the resolving power of MEKC is less than that of conventional chromatography, which is represented by the top curve (t_{MC}/t_0 = infinity). This is true if only f_k is considered. For an elution range value of 3, $f_{k,opt}$ attains 27% of the maximal contribution (at infinite $\tilde{k}'$ and infinite t_{MC}/t_0). For a similar capacity factor (1.73), about 2.4 times as much, 63% is obtained in liquid chromatography. This difference however is easily offset by the superior efficiency of MEKC which also contributes to resolution (**Table 11.1**).

Table 11.1 Relative contributions of efficiency to resolution.

	N	N/m [a]	$\sqrt{N}$
LC	8000	53000	89
	10000	66000	100
	15000	100000	122
MEKC	50000	-	223
	100000	-	316
	200000	-	447

a) For a standard 15 cm analytical LC column

[1] *Contrary to literature practice, we prefer the use of t_{MC}/t_0 to t_0/t_{MC}. The relation of the first to elution range and peak capacity is more obvious than the effect of the latter in eq.(11.2) or (11.5).*

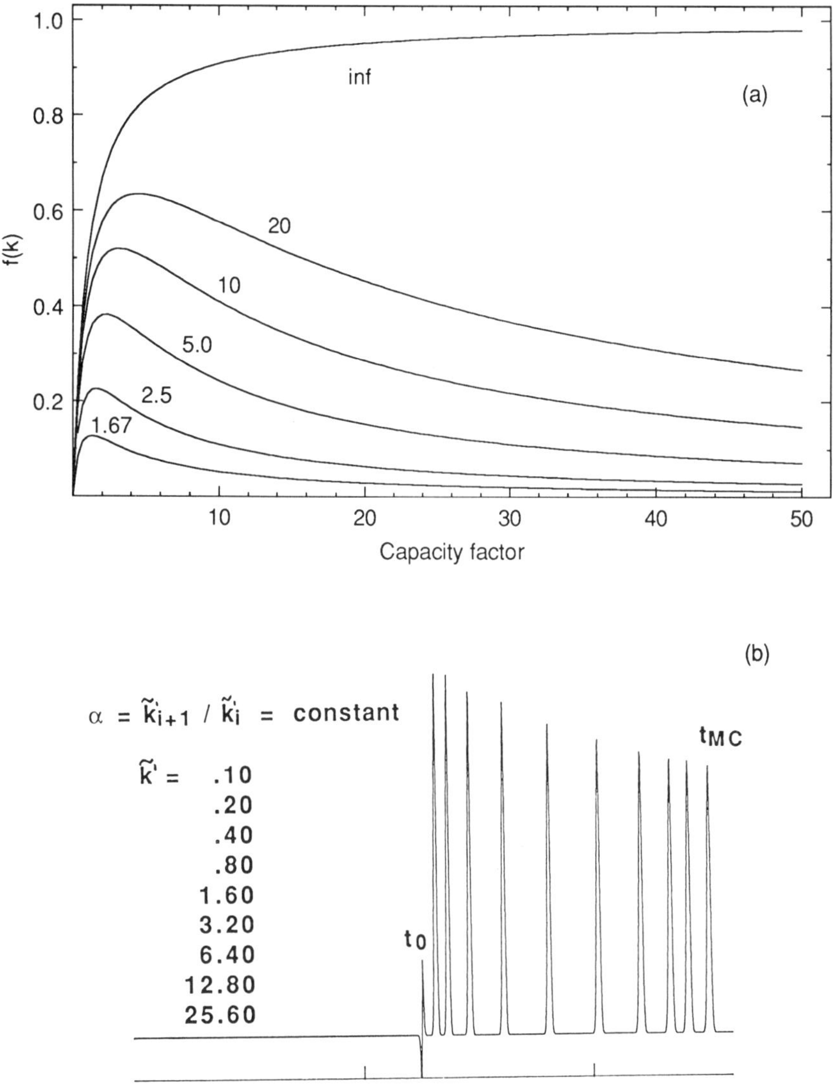

Figure 11.1
(a) Contribution of f_k to resolution as a function of the capacity factor for different elution range values (t_{MC}/t_0).

(b) Simulated chromatogram, ($t_{MC}/t_0 = 2.0$, $\tilde{k}'_{opt} = 1.41$).

Assuming that conditions are selected to obtain optimal efficiency and that selectivity is determined by the type of surfactant, f_k is the most important parameter for resolution optimization. Two strategies can be considered. The first one is $\tilde{k}'$ optimization. The capacity factor is adjusted to obtain the maximal contribution from f_k according to eq.(11.5). A second possibility is elution range optimization. The ratio t_{MC}/t_0 is increased until sufficient resolution is obtained.

11.2 Optimization of the Capacity Factor

The capacity factor is determined by the thermodynamics of the distribution between the two phases (K_{eq}) and by the phase ratio β

$$\tilde{k}' = \beta\, K_{eq} \tag{11.8}$$

In conventional chromatography capacity factors are adjusted through manipulation of K_{eq} by changing the eluting strength of the mobile phase, as the phase ratio is a fixed property of the selected column. In MEKC however it is easy to modify the phase ratio β, simply by changing the surfactant concentration (using ω and ϖ to represent volume fraction and specific partial molar volume respectively) [2] :

$$\beta = \frac{V_{MC}}{V_{aq}} = \frac{\omega_{MC}}{1 - \omega_{MC}} = \frac{(c_{surf} - CMC)\,\varpi_{MC}}{1 - (c_{surf} - CMC)\,\varpi_{MC}} \tag{11.9}$$

With $\varpi_{MC} \approx 0.25$ l/mol (for SDS, 23°C, [3]), the denominator is approximately equal to 1 for a surfactant concentration less than 150 mM, and

$$\tilde{k}' = (c_{surf} - CMC)\,\varpi_{MC}\, K_{eq} \tag{11.10}$$

Capacity factors are linearly correlated with the surfactant concentration as has been confirmed experimentally [2,4] (**Figure 11.2**).

The effect of the phase ratio on resolution has been visualized by a series of simulated chromatograms [1] in **Figure 11.3**. It is obvious that, depending upon the

[1] In this and other simulations in this chapter, the reference elution range (equal to 2) and efficiency (50,000 theoretical plates) have been selected for clarity, giving a somewhat underestimating impression of peak capacity.

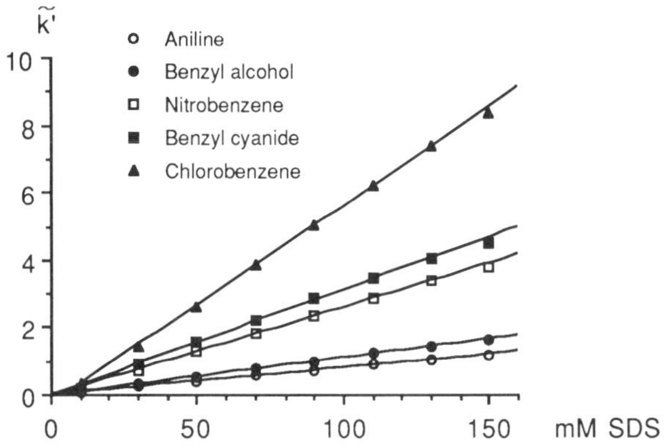

Figure 11.2
Linear dependence of the capacity factor on the surfactant concentration.
Column : 75 µm x 50/57 cm (liquid cooled). Buffer : SDS in 50 mM borate (pH 8.5).
Applied voltage : 20 kV.

initial situation, optimization can require an increase, as well as a decrease in the surfactant concentration.

We have preferred to use simulated chromatograms to exemplify optimization principles. In reality, the series of chromatograms in **Figure 11.3** will look different. Increasing the surfactant concentration can affect electroosmotic mobility (and consequently the elution range) through changes in viscosity and ionic strength. Also, the electrophoretic mobility of the micelles themselves is reduced [5].

Although it has been reported that the electroosmotic mobility remained constant with SDS concentrations between 25 and 150 mM, it was acknowledged that this resulted from compensation by the decreased viscosity, caused by increased current and in-column heating [2]. In thermoregulated capillaries, a dependence of the electroosmotic mobility on surfactant concentration can be observed (**Figure 11.4**). Accordingly, the elution range is increased as also reported by others [6]. The effect of the increased elution range is discussed in the next section.

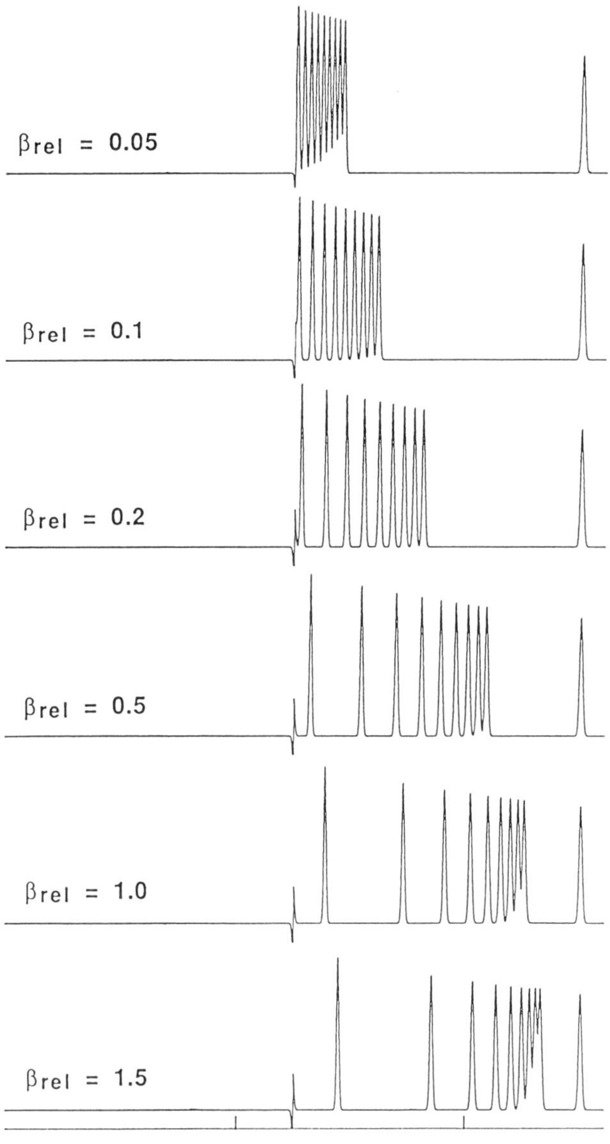

Figure 11.3
Influence of the phase ratio on resolution. The capacity factors of the reference chromatogram (β_{rel} = 1) are as in **Figure 5.2** (simulated chromatograms).

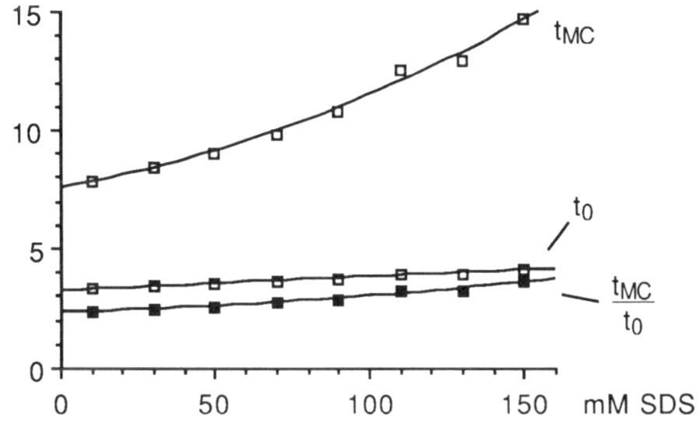

Figure 11.4
Influence of surfactant concentration on t_0 (min), t_{MC} (min) and the elution range t_{MC}/t_0. Electroosmosis and micelle migration time were monitored with formamide and Sudan III respectively. Column : 75 µm x 50/57 cm (liquid cooled). Buffer : SDS in 50 mM borate (pH 8.5). Applied voltage : 20 kV.

11.3 Optimization of the Elution Range

The elution range is determined by the electroosmotic flow and the electrokinetic properties of the micelles. Each of these two factors can be manipulated to optimize the elution range, the first being more practical.

11.3.1 Influence of the Electroosmotic Flow on the Elution Range

The fact that the elution range can be extended, i.e. t_{MC}/t_0 increased, by reducing the electroosmotic flow, i.e. increasing t_0, is not surprising if it is recalled that the electroosmotic flow also affects the migration of the micelles (**Figure 11.5**) :

$$\frac{t_{MC}}{t_0} = \frac{\frac{1}{v_{MC}}}{\frac{1}{v_{eo}}} = \frac{v_{eo}}{v_{MC}} = \frac{v_{eo}}{v_{eo} - v_{ep,MC}} \tag{11.11}$$

The effect of decreasing the electroosmotic flow on the elution range, resolution and peak capacity is obvious from the simulated chromatograms in **Figure 11.6**. Unfortunetaly, the gain in resolving power is accompanied by a strong increase in analysis time. This is also clear if the f_k function (**Figure 11.1a**) is redrawn as a function of the retention time (**Figure 11.7**).

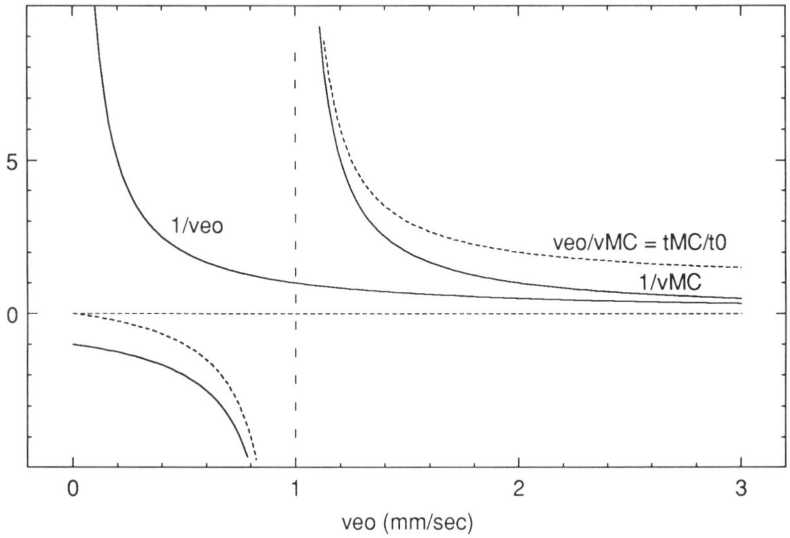

Figure 11.5
Effect of the electroosmotic flow on the elution range ($v_{ep,MC}$ is fixed at 1.0 mm/sec).

Reported values for the elution range of SDS at approximately neutral pH vary between 2.8 and 4.6 [2,6-10]. The wide range is due to variability in the electroosmotic flow, ionic strength and SDS concentration. In a systematic study, starting from an elution range of 4.5 at a pH value of 7.0, infinity, i.e. a true stationary phase, was approached when the pH was lowered to 5 [7]. It has been demonstrated that the electrophoretic mobility of the micelles is virtually independent of the pH [7]. Changes in t_{MC} must be ascribed to changes in the electroosmotic flow (**Figure 11.8**).

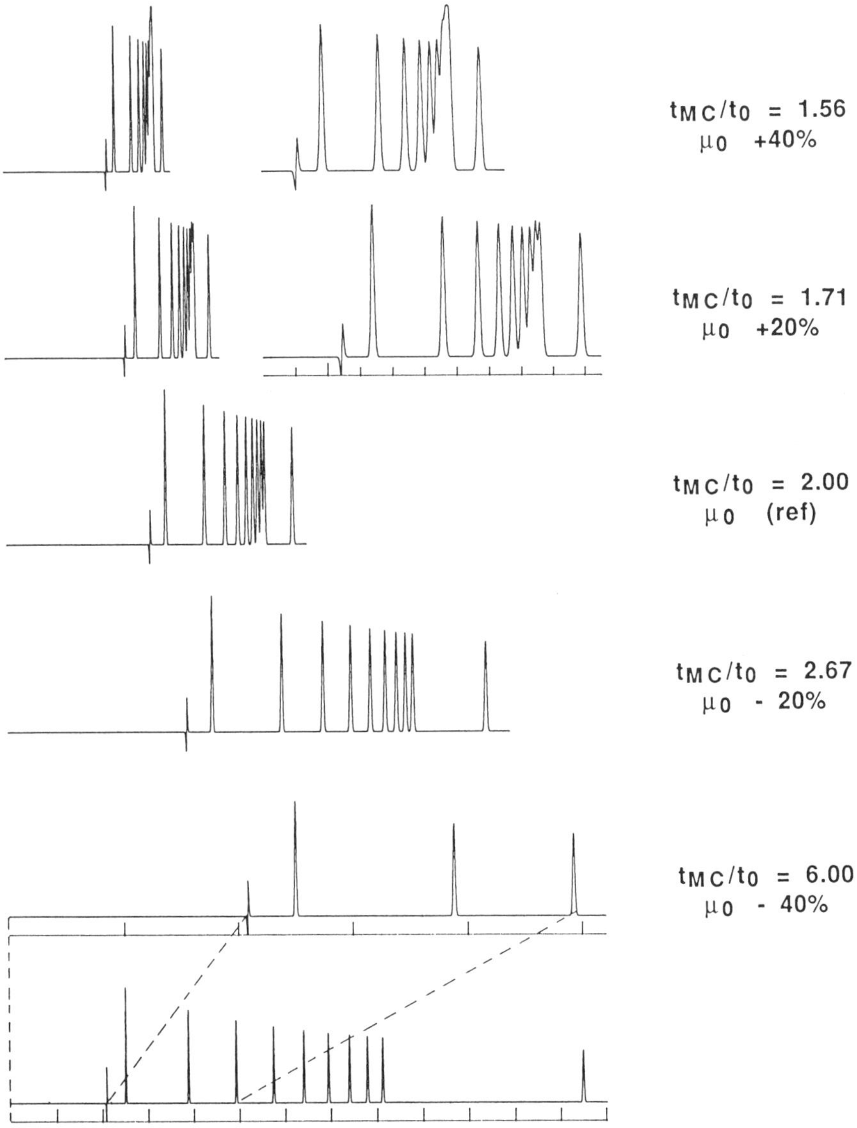

Figure 11.6
Influence of the electroosmotic flow on the elution range and resolution. Capacity factors are as in **Figure 5.2** (simulated chromatograms).

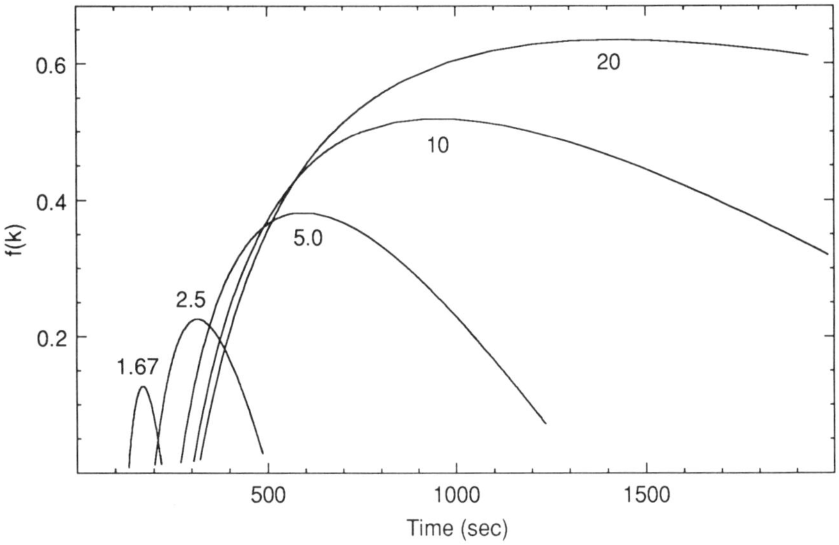

Figure 11.7
Contribution of f_k to the resolution as a function of the retention time for different elution range values. (Column length = 50 cm, $v_{ep,MC}$ fixed at 1.5 mm/sec, v_{eo} ranging from 1.58 to 3.75 mm/sec).

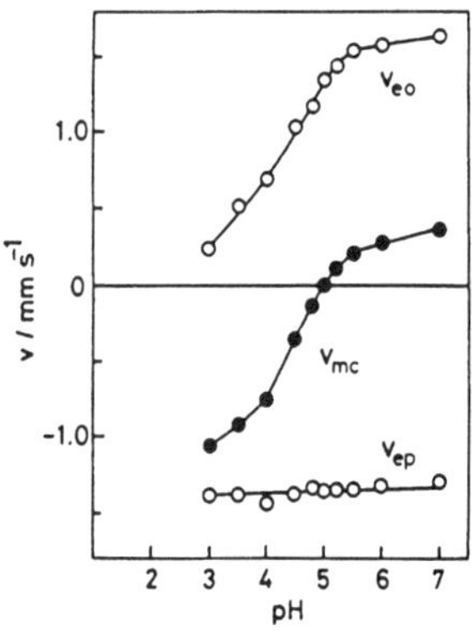

Figure 11.8
Dependence of electroosmotic velocity v_{eo} (methanol), micelle migration velocity v_{mc} (Sudan III) and micelle electrophoretic velocity v_{ep} (calculated) on pH. Column : 50 μm x 50/65 cm. Buffer : 100 mM SDS in 50 mM phosphate. Applied voltage : + or - 15 kV. *(Reprinted with permission [7] © MicroSeparations Inc.))*

Alternatively, the electroosmotic flow can be modified by coating the capillary wall. *Terabe* et al. obtained a pronounced elution range expansion by using a polyethylene glycol coated column (0.2 μm film thickness, Durabond DB-WAX) [11] and SDS as the surfactant. In a methylsilicone coated column (0.2 μm film thickness, Durabond DB-1), however, the electroosmotic flow was increased and the elution range accordingly reduced [11]. This effect has been ascribed to adsorption of SDS on the apolar coating. These results have been confirmed by *Lux* et al. [12]. In TMCS-silanized columns, however, silanol shielding still prevails, leading to an increased elution range (t_{MC}/t_0 estimated at 10) [13].

With some apolar coatings, wall adsorption effects have been observed. A methylsilicone coating caused an elution order reversal between nitrotoluene and toluene, the latter showing a very suspicious loss in efficiency [11]. In TMCS-modified columns, the loss of resolution could be counteracted by adding 10% isopropanol to the mobile phase. [13].

A third way to extend the elution range is the addition of organic modifiers like methanol [14] , isopropanol [13,15] or acetonitrile [14]. Especially with alcohols, the high viscosity of the aqueous mixtures results in long analysis times. It must be noted that the increased viscosity alone cannot explain the enlargement of the elution range, as this influences all mobilities in the same way. Apparently the silica/buffer and micelle/buffer interfaces are affected differerently but the exact nature of those differences has not yet been established. Changes in micelle size and counter-ion association have been hypothized [14].

Only under certain conditions can the use of organic modifiers be considered as a case of elution range optimization. With moderate levels (less than 20% by volume) the effect on the capacity factors and selectivity can only be ignored for the lowest capacity factors ($\tilde{k}' < 1$) [14].

11.3.2 Influence of the Micelle Electrophoretic Mobility on the Elution Range

By increasing the velocity of the micelles, opposite to the electroosmotic flow, the elution range can also be increased (**Figure 11.9**), at least in theory. This is interesting, as the extension of the elution range increases the analysis time less than reduction of the electroosmotic flow. As with organic modifiers, any means of manipulating the electrophoretic properties of micelles will also result in changes in

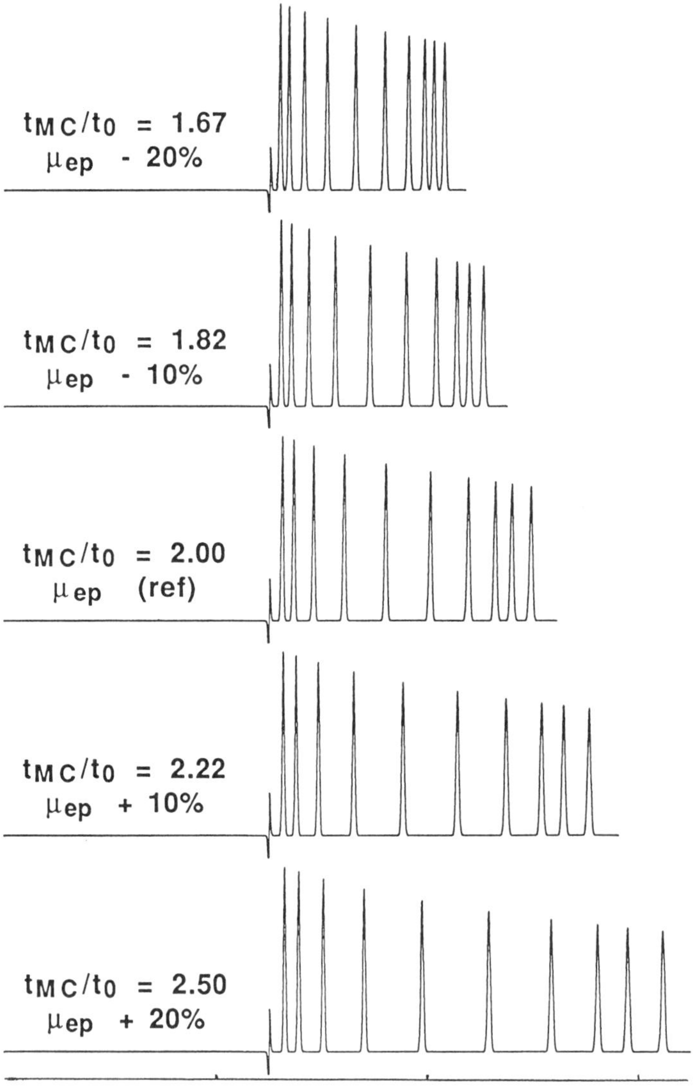

Figure 11.9
Influence of the micelle electrophoretic mobility on the elution range and resolution. Capacity factors are as in **Figure 5.2** (simulated chromatograms).

selectivity.

One method of controlling $\mu_{ep,MC}$ is by varying the chain length of the surfactant. The aggregation number is reduced by shortening the chain (**Table 5.2**) but it is not immediately clear how the reduced effective charge and the reduced micelle size will eventually determine the electrophoretic mobility. A decrease in $\mu_{ep,MC}$ has been reported for the C_{14}-homologue of SDS, resulting in a decrease of the elution range from 3.9 to 3.4 [2]. Accordingly, for the C_{10}-homologue the reverse effect has been observed [6,9]. A similar trend can be noted when the elution range values for a C_{12} and a C_{16} ammonium salt, respectively 2.4 and 2.6 [9], are corrected for the considerable difference in t_0, in which case the C_{12} homologue shows a smaller $\mu_{ep,MC}$. However, due to the importance of the concentration-dependent electroosmotic flow, larger elution ranges (2.6 vs. 2.2) have also been reported for the C_{12} ammonium salt [10].

For a given separation, selecting another surfactant type for elution range optimization will also lead to changes in the capacity factors. For equal surfactant concentrations, the phase ratio will be modified as the CMC's are also different. In many cases, however, changes in the distribution coefficient K_{eq} can also result in selectivity differences. From the few literature data that are available, it can be inferred that, in the case of anionic surfactants, the selectivities of homologues are reasonably well correlated although not invariant (**Figure 11.10**). The addition of two methylene groups seems less influential than changing the surfactant head from sulfate to sulfonate. Small changes in the selectivity, especially when close to 1, can easily offset the effect on resolution that can be expected from changes in the elution range.

A similar correlation has not been observed with cationic surfactants (**Figure 11.11**), even elution order reversals ($\alpha < 1$) are observed on changing the chain length. Also, there is no simple correlation between the selectivity of similar anionic and cationic long hydrophobic chain surfactants. In those cases were a comparison has been made [8-10], elution order reversals are frequently observed. For anionic surfactants, a good correlation between the capacity factor and dodecane-water distribution coefficients has been observed for some compounds [4], indicating the importance of hydrophobic partition. The conjugated π-electron orbitals that are often present can interact with the cationic heads of the surfactant, resulting in different solubilization mechanisms and thus selectivity changes. For ions, attraction or repulsion between solute and micelle will of course be decisive in determining the elution behaviour.

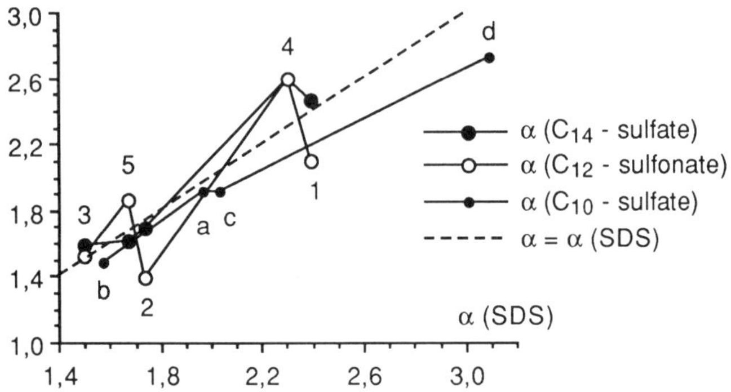

Figure 11.10
Changes in selectivity with surfactant. Sodium decyl sulfate, sodium dodecyl sulfonate and sodium tetradecyl sulfate versus sodium dodecyl sulfate

(1-5 based on $\tilde{k}'$-data from [2], a-d based on α-data from [6]).
1 = Phenol / Resorcinol, 2 = p-Nitroaniline / Phenol, 3 = Nitrobenzene / p-Nitroaniline, 4 = Toluene / Nitrobenzene, 5 = 2-Naphthol / Toluene.
a = Benzene / Benzyl alcohol, b = Acetophenone / Benzene, c = Methyl benzoate / Acetophenone, d = Dimethyl terephthalate / Methyl benzoate.

Alternatively, micelle properties can be changed by using surfactant mixtures. In that case mixed micelles are formed with a CMC intermediate between the respective CMC's of the components in the mixture [16]. *Wallingford* and *Ewing* used mixtures of SDS with sodium octyl sulfate for the separation of borate-complexed catechols [17], claiming increased selectivity and resolution, but their data do not allow the evaluation of relative contributions from selectivity and elution range. *Balchunas* and *Sepaniak* used Triton-X-100[1] / SDS-mixtures for the analysis of derivatized amines [15]. Their mobile phase also contained isopropanol, and it is not clear whether the observed selectivity changes are due to the organic modifier or to the presence of the neutral surfactant. The Triton-X-100, being neutral, reduces the charge of the micelle, and consequently its electrophoretic mobility, and was used to compensate for the elution range extension caused by the isopropanol. *Rasmussen* et al. used SDS/Brij-35[2] to resolve benzaldehyde and benzene [6], and

1 Triton®-X-100 = polyoxyethylene p-isooctylphenyl ether.
2 Brij® 35 = polyoxyethylene(23) lauryl ether.

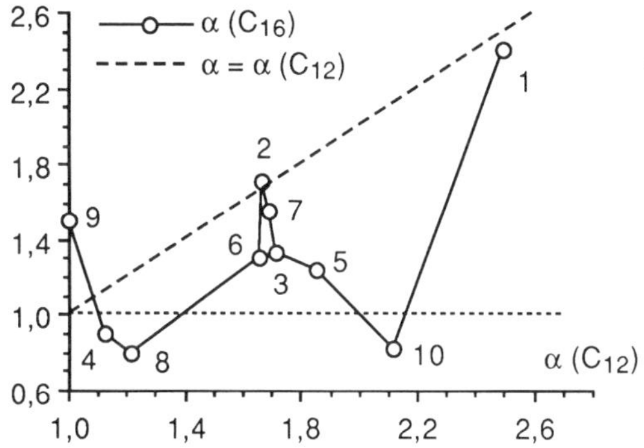

Figure 11.11
Changes in selectivity with surfactant. Dodecyltrimethylammonium bromide versus hexadecyltrimethylammonium bromide (based on $\tilde{k}'$-data from [9]).
1 = Nitrobenzene / Aniline, 2 = Phenol / Nitrobenzene, 3 = Toluene / Phenol, 4 = 2,4-Dinitrotoluene / Toluene, 5 = Benzidine / 2,4-Dinitrotoluene, 6 = 1-Aminonaphthalene / Benzidine, 7 = Naphthalene / 1-Aminonaphthalene, 8 = 1-Nitronaphthalene / Naphthalene, 9 = Anthracene / 1-Nitronaphthalene, 10 = 1-Naphthol / Anthracene.

although the separation was ascribed to changes in selectivity, they could reduce t_{MC} and analysis time, in a way comparable to Figure 11.10 (reading from bottom to top), by increasing the Brij-35 to SDS ratio [18].

11.3.3 Peak Capacity

Peak capacity (n) is defined as the number of peaks that will fit in a particular elution time interval, assuming unit resolution between each pair of successive peaks

$$n = \int_{t_1}^{t_2} \frac{\partial t}{4\,\sigma_t} \tag{11.12}$$

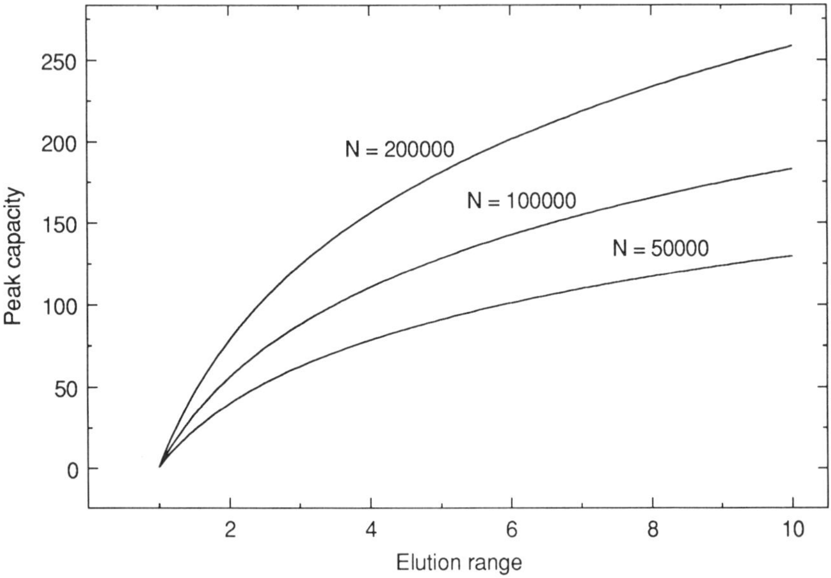

Figure 11.12
Peak capacity as a function of efficiency and elution range (eq.(11.13)).

In chromatography, this interval is rather arbitrarily chosen but in MEKC it can be very logically delimited by t_0 and t_{MC} [2] (**Figure 11.12**) :

$$n = \frac{\sqrt{N}}{4} . \ln\left(\frac{t_{MC}}{t_0}\right) \qquad (11.13)$$

Resolution optimization methods, based on elution range extension, will also increase peak capacity. In eq.(11.13), it is assumed that efficiency is constant along the time axis. If efficiency is mainly determined by axial diffusion[1] this condition is not true :

$$N = \frac{t_R^2}{\sigma_t^2} = \frac{L^2}{\sigma_L^2} = \frac{L^2}{2\, D_m\, t_R} \qquad (11.14)$$

[1] *To simplify integration, the difference between D_m and $D_{m,MC}$ in eq.(10.1) is ignored.*

from which it can be derived that :

$$n = \frac{L}{2\sqrt{2 D_m}} \cdot \left(\frac{\sqrt{\frac{t_{MC}}{t_0}} - 1}{\sqrt{t_{MC}}} \right) \qquad (11.15)$$

Assuming that $L = 0.5$ m, $t_{MC} = 1{,}140$ sec and $t_0 = 300$ sec and $D_m = 10^{-9}\ m^2$ / sec, eq. (11.15) leads to $n = 157$. Eq.(11.13) would yield $n = 215$, using $N = N_0 = 416{,}670$ (overestimating average N) or $n = 111$, using $N = N_{MC} = 109{,}650$ (underestimating average N).

11.4 Alternative Elution Modes

Until now, it was assumed that all neutral molecules elute within the elution range, defined by t_0 and t_{MC}. In **Figure 11.5** it was already suggested that the electroosmotic flow could be lowered to such a level that the micelles move in the other direction [7,19,20] (**Figure 11.13**). In that case they move away from the detector, resulting in a negative, virtual, t_{MC} value. According to eq (5.3), and considering that t_{MC}/t_0 is negative, t_R will remain positive as long as

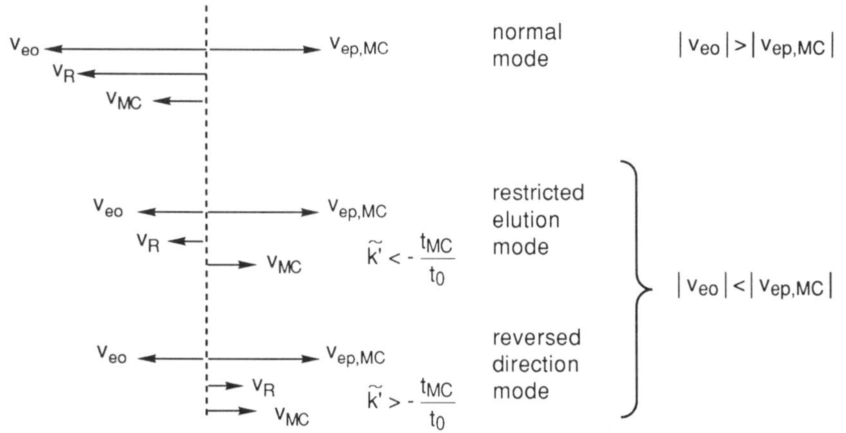

Figure 11.13
Elution modes in Micellar Electrokinetic Chromatography.

$$\widetilde{k'} < -\frac{t_{MC}}{t_0} \qquad (11.16)$$

The elution of all components in a mixture is no longer guaranteed [11,21,22], therefore we call this the *restricted elution mode*. Compounds with a larger capacity factor are carried along with the micelles in the opposite direction and will require opposite polarity to be detected : *reversed direction mode* (**Figure 11.14**). In this mode, more hydrophobic analytes will elute first [23-26].

In alternative elution modes, t_0 and t_{MC} have opposite signs so that two measurements are necessary for the determination of capacity factors, one at each polarity [21,22].

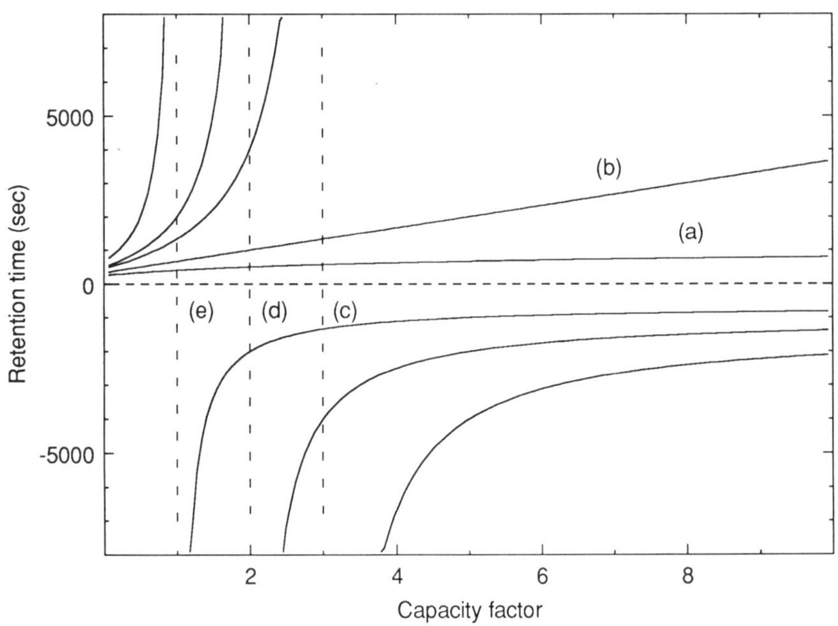

Figure 11.14
Calculated retention times (Column length 50 cm, $v_{ep,MC}$ = -1.5 mm/sec)
a) v_{eo} = 2 mm/sec or t_{MC}/t_0 = 4 (normal mode)
b) v_{eo} = 1.5 mm/sec or t_{MC}/t_0 = infinity (normal mode, borderline)
c) v_{eo} = 1.125 mm/sec or t_{MC}/t_0 = -3
d) v_{eo} = 1 mm/sec or t_{MC}/t_0 = -2
e) v_{eo} = 0.75 mm/sec or t_{MC}/t_0 = -1
c)-e) : restricted elution mode if $t_R > 0$, reversed direction mode if $t_R < 0$

In normal mode, the contribution from f_k to resolution is limited to 1 (at infinite capacity factor and elution range). In the other modes, however, a much larger contribution is possible (**Figure 11.15**), perhaps at the cost of increased analysis time.

The reduction in electroosmotic flow necessary to realize the alternative elution modes can be obtained by reducing the buffer pH, coating the column or the addition of methanol to the buffer. With SDS as the surfactant, restricted elution mode is observed at pH values below 5 [7] and reversed direction mode has been used with SDS and with bile salts at pH values between 3 and 4 [7,23-26]. Unfortunately, this is also the pH range in which the electroosmotic flow is most susceptible to hysteresis effects [27], resulting in loss of reproducibility [7]. Restricted migration mode has been observed in polyethylene glycol coated fused silica columns [11]. At a pH value of 8, and with SDS, normal elution mode stops when the methanol content of the buffer is increased to ca. 20 % by volume [21,22].

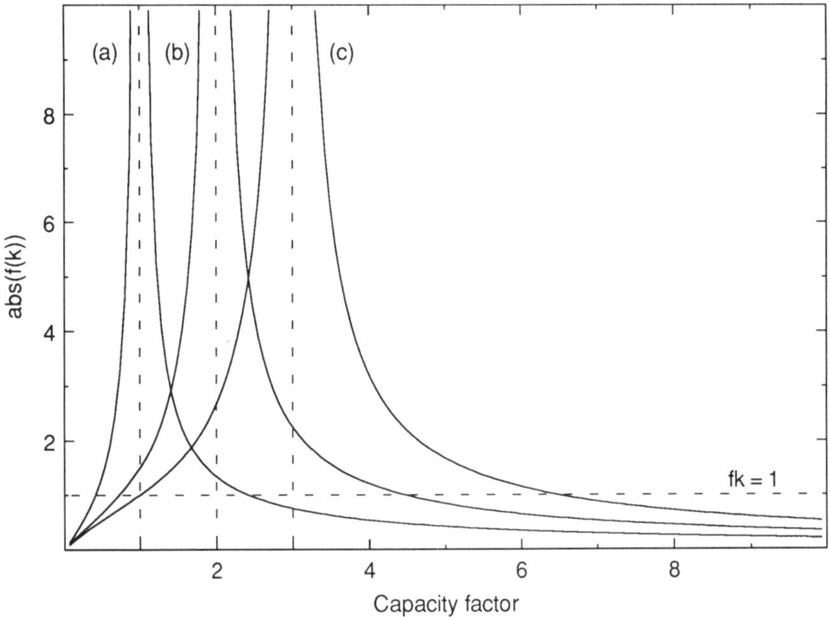

Figure 11.15
Influence of capacity factor on resolution in alternate elution modes compared with normal mode (f_k = 1). a) t_{MC}/t_0 = -1, b) t_{MC}/t_0 = -2, c) t_{MC}/t_0 = -3
The left side of each curve represents restricted elution mode, the right side represents reversed direction mode.

11.5 The MEKC Analysis of Hydrophobic Compounds

11.5.1 Strategies for the Analysis of Hydrophobic Compounds

From the discussion on the resolution function in MEKC, it follows that resolution will be optimal for hydrophilic compounds. Hydrophobic compounds will elute close together, with little resolution, as the capacity factor is increased along with micellar solubilization. Several methods have been developed to reduce capacity factors. The use of bile salts instead of SDS has been proposed [28], as these bile salts show a relatively small solubilizing power. But the fundamental contradiction between hydrophobicity and low capacity factors remains. If a high surfactant concentration is needed for solubilizing the hydrophobic compounds, high capacity factors can result and, as a consequence, low resolution is obtained.

Alternatively, the polar nature of the aqueous phase can be reduced. *Terabe* et al. separated polyaromatic hydrocarbons, chlorinated benzenes and PCB's by adding cyclodextrins to the micellar buffer solution [29,30] (CD-MEKC = cyclodextrin modified MEKC). Partition between aqueous and micellar phase is shifted by inclusion of the analytes in the cyclodextrin cavity (**Figure 11.16**). The importance of the goodness of fit between analyte and cavity is demonstrated by the different effects of β- and γ-cyclodextrin on separation [29].

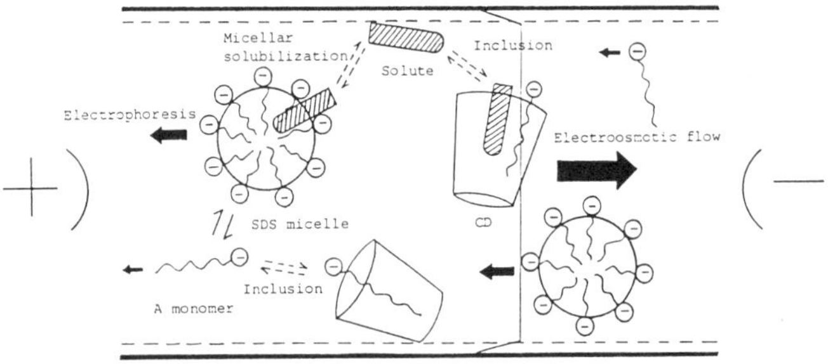

Figure 11.16
Micellar solubilization and cyclodextrin inclusion in cyclodextrin-modified MEKC. *(Reprinted with permission [31] © Elsevier Science Publishers B.V.)*

Organic modifiers can also be added to the aqueous phase : methanol [14,21,22], acetonitrile [14,32] or urea [33]. The use of these organic modifiers for controlling the elution range has already been discussed. With hydrophobic compounds, the relative effects of the modified elution range and changed capacity factors is difficult to predict. Due to increased buffer complexity, variables like t_0, t_{MC} and $\tilde{k}'$ cannot be independently controlled. The neat division between elution range optimization and $\tilde{k}'$ optimization, which already shows, some mutual interference, is further disturbed by the presence of organic modifier. Depending upon its concentration, everything is liable to change : electroosmotic flow, micelle properties and distribution coefficients (**Figure 11.17**).

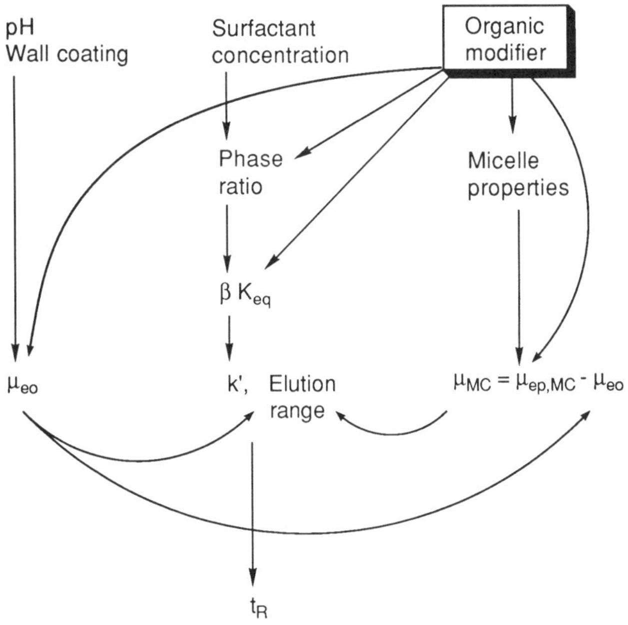

Figure 11.17
Effect of the organic modifier on electrokinetic properties in a micellar system.

Analytical approaches to obtain optimal resolution per unit of time have been described [20,34] but, apart from the many generalizations used, application in resolution optimization requires a rigid control of the characteristics of the system. If this is not possible, which is currently the case with organic modifiers, a global approach is necessary, combining all possible optimization modes and also taking into account changes in N and α. Given the many parameters that can be varied, the complete resolution response surface is a complicated multidimensional structure whose complete elucidation that would require a tremendous experimental effort. *Vindevogel* and *Sandra* have used a statistical design method to circumvent this problem and considered 5 buffer parameters in only 8 experiments [32] (**Table 11.2** and **11.3**, **Figure 11.18**) for the resolution optimization of four testosterone esters. The differences in resolution, analysis time and chromatogram quality in **Figure 11.18**, obtained with relatively small changes in buffer composition, illustrate the problem of resolution optimization when high levels of organic modifier are present.

Table 11.2 Buffers used for the resolution optimization of four testosterone esters according to a Plackett-Burman design. *(Reprinted with permission [32] © American Chemical Society)*

	dummy1 a)	pH b)	%shs c)	%acn d)	mMsurf e)	dummy2 a)	mMbuf f)
Buffer#1	+0	9	10	40	50	-0	20
Buffer#2	-0	9	10	50	40	+0	20
Buffer#3	-0	8	10	50	50	-0	40
Buffer#4	+0	8	0	50	50	+0	20
Buffer#5	-0	9	0	40	50	+0	40
Buffer#6	+0	8	10	40	40	+0	40
Buffer#7	+0	9	0	50	40	-0	40
Buffer#8	-0	8	0	40	40	-0	20

a) Dummy1 and dummy2 are used for estimating the system variability.
b) Indicative, not thermodynamic value.
c) Part of the surfactant was replaced by sodium heptyl sulfate (% of molar concentration)
d) Acetonitrile content (volume percent)
e) Concentration of surfactant (SDS)
f) Buffer concentration (borate)

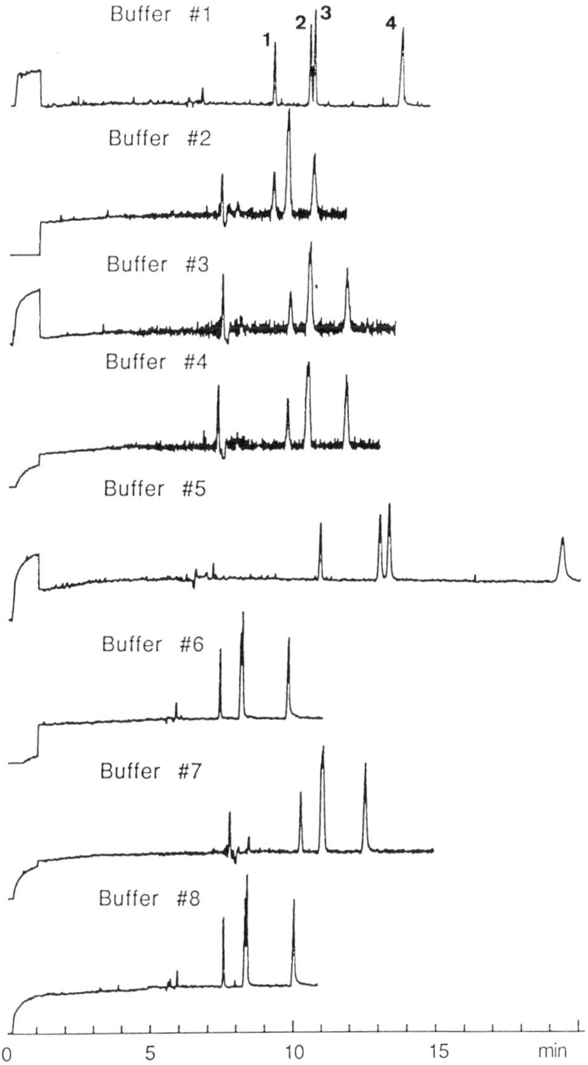

Figure 11.18
Resolution optimization for the MEKC analysis of testosterone esters : eight chromatograms, according to the buffer compositions in **Table 11.2**.

Capillary : 75 µm x 50/57 cm. Applied voltage : 25 kV. Detection : 254 nm.
1 = Testosterone propionate , 2 = Testosterone phenylpropionate, 3 = Testosterone isocaproate, 4 = Testosterone decanoate. *(Reprinted with permission [32] © American Chemical Society)*

Table 11.3 Effects of buffer parameters (see **Table 11.2**) on electroosmotic migration time (t_0), migration time of last-eluting testosterone ester (t_{T4}), noise, current, resolution of the critical pair (R_S) and the efficiency of the first eluting testosterone ester (N_{T1}). Only significant effects are shown. *(Reprinted with permission [32] © American Chemical Society)*

	t_0 (min)	t_{T4} (min)	Noise (μAU)	Current (μA)	R_S	N_{T1}
dummy1						
pH	0.50 a)	3.21		15	0.38	
%shs	-0.16	-1.96			-0.28	
%acn	1.49		67		-0.77	-78025
mMsurf	0.27	3.42		19	0.53	
dummy2						
mMbuf	0.20				0.18	
Range b)	2.21	9.59	116	36	1.76	134700
σ c)	0.05	0.95	20	4	0.03	25600

a) This unexpected result is related to the increase of the ionic strength with pH and has already been discussed (**3.3**).
b) Difference between highest and lowest value found.
c) Standard deviation as measured from dummy1 and dummy2

11.5.2 The Separation Mechanism in Hydro-Organic Solvents

The presence of the organic modifier poses the problem of micelle integrity and its consequences for the separation mechanism. The breakdown of micelles by organic modifiers is not a matter of principle but originates from the balance between several processes. The driving force for micelle formation [35] is the energy gained by exclusion of the hydrophobic chains from the polar solvent matrix. The counteracting factor is the accumulation of charges on the micellar surface. In the case of alcohols, these effects have been described. A decrease in the CMC of cationic [36] and anionic [37] surfactants has been observed on the addition of short-chain alcohols. This has been explained by a decrease in the surface charge density by inclusion of the neutral organic modifier in the micelle outer sphere [37]. The linear relationship between the CMC decrease and the molarity of the alcohol has been shown for ethanol concentrations up to 10%. At higher concentrations, as the energy gain by exclusion becomes less important, the CMC decreases again, until finally micelles disappear.

Similar detailed studies do not exist for acetonitrile. Although the existence of micelles at a high concentration of acetonitrile has not been unequivocally demonstrated, electrokinetic separations of neutral molecules have been reported in such a medium [32,38]. Hydrophobic association has been proposed as the separation mechanism in a system consisting of tetrahexylammonium salts in water-acetonitrile mixtures [38]. In that case, attraction between analyte π-electrons and the ammonium salts is an alternative explanation. This type of attraction is also involved in the interaction between aromatic compounds and the charged outer sphere of cationic micelles [39] but cannot be considered with anionic surfactants. With SDS, the dependence of mobilities on surfactant concentration strongly suggests that a threshold surfactant concentration is necessary for separation (**Figure 11.19**) [32].

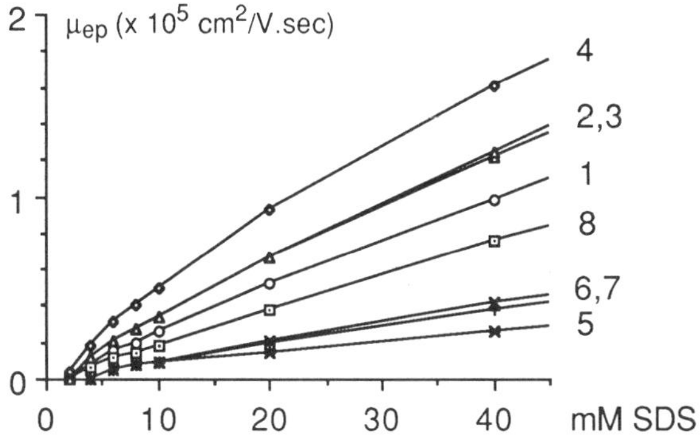

Figure 11.19
Influence of the surfactant concentration on the apparent electrophoretic mobility. The correction for electroosmosis is based on the migration time of formamide. Buffer : 40% CH_3CN, 7.5 mM borax. 1 = Testosterone propionate, 2 = Testosterone phenylpropionate, 3 = Testosterone isocaproate, 4 = Testosterone decanoate, 5 = Benzyl alcohol, 6 = Phenylacetone, 7 = Benzyl cyanide, 8 = Triton-X-100. *(Reprinted with permission [32] © American Chemical Society)*

11.5.3 Gradient Elution in MEKC

The strong interdependence between resolution and the capacity factor poses a problem for the analysis of complex mixtures. In that case, sufficient resolution must be available over a wide range of capacity factors. Temperature programming, based on the observed reduction in $\tilde{k}'$ with increased temperature [2], has been explored [15] but is obviously limited by the strong current increase, especially at high surfactant concentrations.

In initial attempts at solvent programming [15], step-wise gradients were created by adding portions of a stronger eluting buffer to the inlet buffer reservoir, which required the applied field to be turned off momentarily. With this method, using a combined gradient of isopropanol and Triton-X-100, the possibility of gradient elution was demonstrated. By moving the high-voltage side to the outlet reservoir, maintaining the inlet at ground, motor-driven syringes can be used to create continuous gradients of various shapes [40]. The inlet reservoir is continuously stirred and serves as a mixing chamber. This method has been applied to the generation of acetonitrile gradients (**Figure 11.20**) [40-42].

Solvent programming is a complex phenomenon in MEKC. Elution range, electroosmotic flow and selectivity change during gradient development. A computational model has been developed to predict retention times and peak resolution as a function of the gradient, based on isocratic elution data [41].

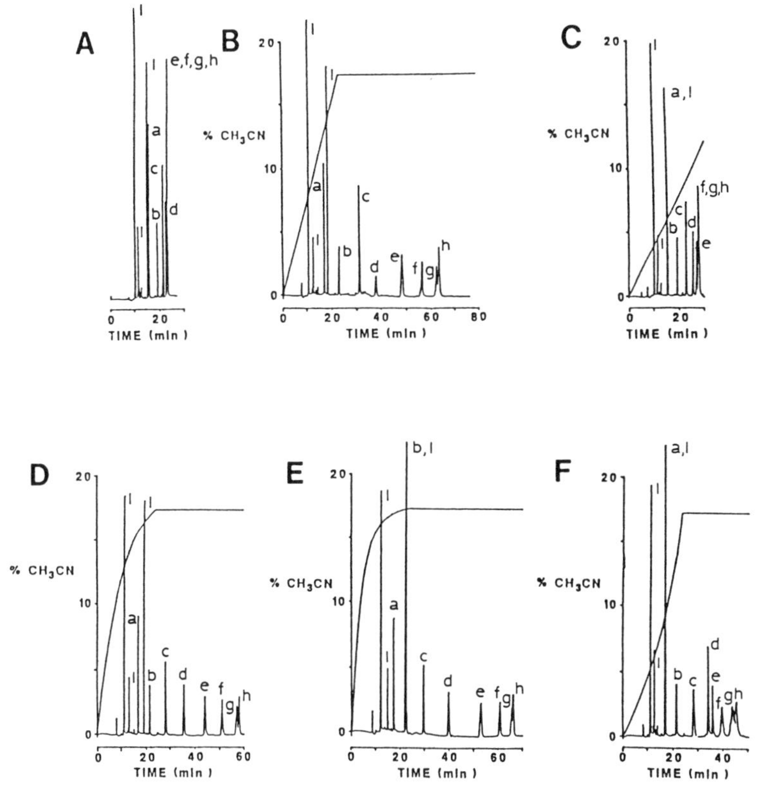

Figure 11.20
MEKC separation of NBD-derivatized amines with solvent programming and laser-induced fluorescence detection. Buffer : 50 mM SDS in phosphate/borax.
a = NBD-n-propylamine, b = NBD-n-butylamine, c = NBD-n-pentylamine, d = NBD-n-hexylamine, e = NBD-n-heptylamine, f = NBD-n-octylamine, g = NBD-n-decylamine, h = NBD-n-dodecylylamine, I = impurities.
A = without organic modifier, B,C = linear acetonitrile gradients, D,E = convex acetonitrile gradients, F = a concave acetonitrile gradient.
(Reprinted with permission [40] © Elsevier Science Publishers B.V.)

References

[1] P.Sandra, J.High Resolut.Chromatogr.Chromatogr.Commun., **12** (1989) 82.
Resolution - Definition and nomenclature.
[2] S.Terabe, K.Otsuka, T.Ando, Anal.Chem., **57** (1985) 834.
Electrokinetic chromatography with micellar solution and open-tubular capillary.
[3] P.Mukerjee, J.Phys.Chem., **66** (1962) 1733.
The partial specific volume and the density of micelles of association colloidal electrolytes.
[4] K.Saitoh, C.Kiyohara, N.Suzuki, J.High Resolut.Chromatogr., **14** (1991) 245.
Mobilities of metal-β-diketonate complexes in micellar electrokinetic chromatography.
[5] D.Stigter, K.J.Mysels, J.Phys.Chem., **59** (1955) 45.
Tracer electrophoresis. II. The mobility of the micelle of sodium lauryl sulfate and its interpretation in terms of zeta potential and charge.
[6] H.T.Rasmussen, L.K.Goebel, H.M.McNair, J.Chromatogr., **517** (1990) 549.
Micellar electrokinetic chromatography employing sodium alkyl sulfates and Brij 35.
[7] K.Otsuka, S.Terabe, J.Microcol.Sep., **1** (1989) 150.
Effect of pH on electrokinetic velocities in micellar electrokinetic chromatography.
[8] K.Otsuka, S.Terabe, T.Ando, J.Chromatogr., **332** (1985) 219.
Electrokinetic chromatography with micellar solutions. Separation of phenylthiohydantoin-amino acids.
[9] D.E.Burton, M.J.Sepaniak, M.P.Maskarinec, J.Chromatogr.Sci., **25** (1987) 514.
Evaluation of the use of various surfactants in micellar electrokinetic capillary chromatography.
[10] J.Liu, F.Banks, M.Novotny, J.Microcol.Sep., **1** (1989) 136.
High-speed micellar electrokinetic capillary chromatography of the common phosphorylated nucleosides.
[11] S.Terabe, H.Utsumi, K.Otsuka, T.Ando, T.Inomata, S.Kuze, Y.Hanaoka, J.High Resolut. Chromatogr.Chromatogr.Commun., **9** (1986) 666.
Factors controlling electroosmotic flow in open-tubular capillaries in electrokinetic chromatography.
[12] J.A.Lux, H.Yin, G.Schomburg, J.High Resolut.Chromatogr., **13** (1990) 145.
Influence of polymer coating of capillary surfaces on migration behavior in micellar electrokinetic capillary chromatography.
[13] A.T.Balchunas, M.J.Sepaniak, Anal.Chem., **59** (1987) 1466.
Extension of elution range in micellar electrokinetic capillary chromatography.
[14] J.Gorse, A.T.Balchunas, D.F.Swaile, M.J.Sepaniak, J.High Resolut.Chromatogr.Chromatogr. Commun., **11** (1988) 554.
Effects of organic mobile phase modifiers in micellar electrokinetic capillary chromatography.
[15] A.T.Balchunas, M.J.Sepaniak, Anal.Chem., **60** (1988) 617.
Gradient elution for micellar electrokinetic capillary chromatography.
[16] K.Shinoda, J.Phys.Chem., **58** (1954) 541.
The critical micelle concentration of soap mixtures (two-component mixture).
[17] R.A.Wallingford, P.D.Curry, A.G.Ewing, J.Microcol.Sep., **1** (1989) 23.
Retention of catechols in capillary electrophoresis with micellar and mixed micellar solutions.
[18] H.T.Rasmussen, L.K.Goebel, H.M.McNair, J.High Resolut.Chromatogr., **14** (1991) 25.
Optimization of resolution in micellar electrokinetic chromatography.
[19] P.Gareil, Chromatographia, **30** (1990) 195.
Migration time and resolution in micellar electrokinetic capillary chromatography.
[20] K.Ghowsi, J.P.Foley, R.J.Gale, Anal.Chem., **62** (1990) 2714.
Micellar electrokinetic capillary chromatography theory based on electrochemical parameters : Optimization for three modes of operation.
[21] M.M.Bushey, J.W.Jorgenson, J.Microcol.Sep., **1** (1989) 125.
Effects of methanol-modified mobile phase on the separation of isotopically substituted compounds by micellar electrokinetic capillary chromatography.
[22] M.M.Bushey, J.W.Jorgenson, Anal.Chem., **61** (1989) 491.
Separation of dansylated methylamine and dansylated methyl-d_3-amine by micellar electrokinetic capillary chromatography with methanol-modified mobile phase.
[23] K.Otsuka, S.Terabe, J.Chromatogr., **515** (1990) 221.
Enantiomeric resolution by micellar electrokinetic chromatography with chiral surfactants.
[24] S.Terabe, M.Shibata, Y.Miyashita, J.Chromatogr., **480** (1989) 403.
Chiral separation by electrokinetic chromatography with bile salt micelles.

[25] H.T.Rasmussen, H.M.McNair, J.High Resolut.Chromatogr. Chromatogr. Commun., **12** (1989) 635.
Influence of pH on elution behavior in micellar electrokinetic capillary chromatography.
[26] A.D.Tran, T.Blanc, E.Leopold, J.Chromatogr., **516** (1990) 241.
Free solution capillary electrophoresis and micellar electrokinetic resolution of amino acid enantiomers with L- and D-Marfey's reagent.
[27] W.J.Lambert, D.L.Middleton, Anal.Chem., **62** (1990) 1585.
pH hysteresis effect with silica capillaries in capillary zone electrophoresis.
[28] H.Nishi, T.Fukuyama, M.Matsuo, S.Terabe, J.Chromatogr., **513** (1990) 279.
Separation and determination of lipophilic corticosteroids and benzothiazepin analogues by micellar electrokinetic chromatography using bile salts.
[29] S.Terabe, Y.Miyashita, O.Shibata, E.R.Barnhart, L.R.Alexander, D.G.Patterson, B.L.Karger, K.Hosoya, N.Tanaka, J.Chromatogr., **516** (1990) 23.
Separation of highly hydrophobic compounds by cyclodextrin-modified micellar electrokinetic chromatography.
[30] H.Nishi, M.Matsuo, J.Liq.Chromatogr., **14** (1991) 973.
Separation of corticosteroids and aromatic hydrocarbons by cyclodextrin-modified micellar electrokinetic chromatography.
[31] H.Nishi, T.Fukuyama, S.Terabe, J.Chromatogr., **553** (1991) 503.
Chiral separation by cyclodextrin modified micellar electrokinetic chromatography.
[32] J.Vindevogel, P.Sandra, Anal.Chem., **63** (1991) 1530.
Resolution optimization in micellar electrokinetic chromatography : Use of Plackett-Burman statistical design for the analysis of testosterone esters.
[33] S.Terabe, Y.Ishihama, H.Nishi, T.Fukuyama, K.Otsuka, J.Chromatogr, **545** (1991) 359.
Effect of urea addition in micellar electrokinetic chromatography.
[34] J.P.Foley, Anal.Chem., **62** (1990) 1302.
Optimization of micellar electrokinetic chromatography.
[35] H.V.Tartar, J.Phys.Chem., **59** (1955) 1195.
A theory of the structure of the micelles of normal paraffin chain salts in aqueous solution.
[36] S.H.Herzfeld, M.L.Corrin, W.D.Harkins, J.Phys.Colloid Chem., **54** (1950) 271.
The effect of alcohols and of alcohols and salts on the critical micelle concentration of dodecylammonium chloride.
[37] K.Shinoda, J.Phys.Chem., **58** (1954) 1136.
The effect of alcohols on the critical micelle concentrations of fatty acid soaps and the critical micelle concentration of soap mixtures.
[38] Y.Walbroehl, J.W.Jorgenson, Anal.Chem., **58** (1986) 479.
Capillary zone electrophoresis of neutral organic molecules by solvophobic association with tetraalkylammonium ion.
[39] M.G.Khaledi, J.K.Strasters, A.H.Rodgers, E.D.Breyer, Anal.Chem., **62** (1990) 130.
Simultaneous enhancement of separation selectivity and solvent strength in reversed-phase liquid chromatography using micelles in hydro-organic solvents.
[40] M.J.Sepaniak, D.F.Swaile, A.C.Powell, J.Chromatogr., **480** (1989) 185.
Instrumental developments in micellar electrokinetic capillary chromatography.
[41] A.C.Powell, M.J.Sepaniak, J.Microcol.Sep., **2** (1990) 278.
Development of a model for predicting retention times in solvent-gradient micellar electrokinetic capillary chromatography.
[42] M.J.Sepaniak, D.F.Swaile, A.C.Powell, R.O.Cole, J.High Resolut.Chromatogr., **13** (1990) 679.
Capillary electrokinetic separations : influence of mobile phase composition on performance.

12 General Applications

Capillary electrophoretic techniques have recently been included in the review sections of *Analytical Chemistry*, covering fundamental aspects and applications of MEKC. The first review [1] covers the literature between 1983 and 1989. This survey of applications extends this period to fall 1991. Unless otherwise specified, the most commonly used surfactant, SDS, and uncoated fused silica capillaries are assumed.

12.1 Derivatives of Amino Acids, Aliphatic Amines and Amino Sugars

12.1.1 Isoindole Derivatives

Ortho-phthaldialdehyde (OPA) is a well-known, commercially available derivatizing reagent for primary amines and amino acids [2]. Condensation of OPA with the substrate, in the presence of a reducing agent like 2-mercaptoethanol, leads to fluorescing isoindole derivatives. OPA has been used for post-capillary derivatization in the CZE analysis of amino acids [3,4]. The superior separating capability of MEKC over CZE for OPA-derivatized amino acids has been demonstrated by *Liu* et al. [5]. Fourteen of the common amino acids were separated in a mixture which also contained γ-aminobutyric acid, taurine and 2-aminoethanol (used as an internal standard) (**Figure 12.1**). The authors suspected that the well-known instability of the isoindoles was aggravated by in-column heating during analysis. Improved stability was attained on replacing 2-mercaptoethanol with ethanethiol.

Separations of limited sets of NDA-derivatized amino acids, which are more stable [6], have been demonstrated in CZE [7], but the analysis is characterized by many coelutions [8]. MEKC analysis, however, has not yet been reported.

In order to obtain a better match between the excitation maximum and the common laser excitation lines (325 nm and 442 nm for the He/Cd laser and 488 nm for the Argon ion laser), several other isoindole-forming derivatization reagents have been synthesized [9-12] (**Figure 12.2**). As some of these derivatives are too hydrophobic to be used in CZE, 3-(4-carboxybenzoyl)-2-quinolinecarboxaldehyde (CBQCA) has been synthesized to overcome that limitation. The separation and laser-induced fluorescence detection of 17 CBQCA-derivatized amino acids has been realized with 50 mM SDS in a TES-buffer (pH 7.0) [11]. The presence of SDS, although necessary for the separation, increased the detection limit (evaluated at 10-70 attomole) by one order of magnitude, which was ascribed to increased light-scattering. The same reagent can be used for the MEKC separation of amino sugars (**Figure 12.3**) and has also been applied to the analysis of oligo-aminosugars in chitosan and glycoprotein hydrolysates [12].

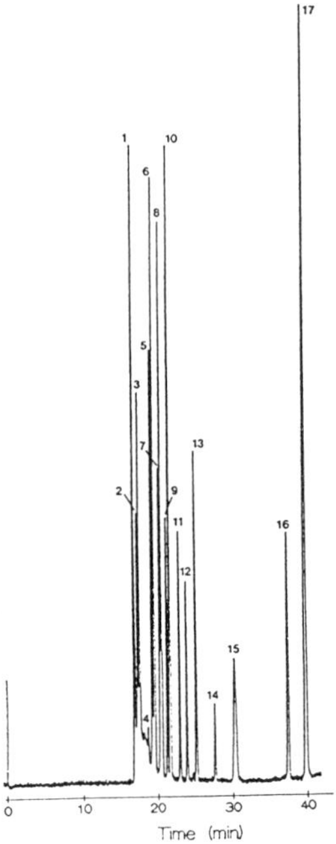

Figure 12.1
MEKC separation and fluorescence detection (λ_{exc} = 365 nm) of OPA/Ethanethiol-derivatized amino acids with an anionic surfactant. Capillary : 50 µm x 53/86 cm. Buffer : 50 mM SDS in 50 mM borate (pH 9.5) with 15% MeOH and 2% THF. Applied voltage : 23 kV. *(Reprinted with permission [5] © Elsevier Science Publishers B.V.)*

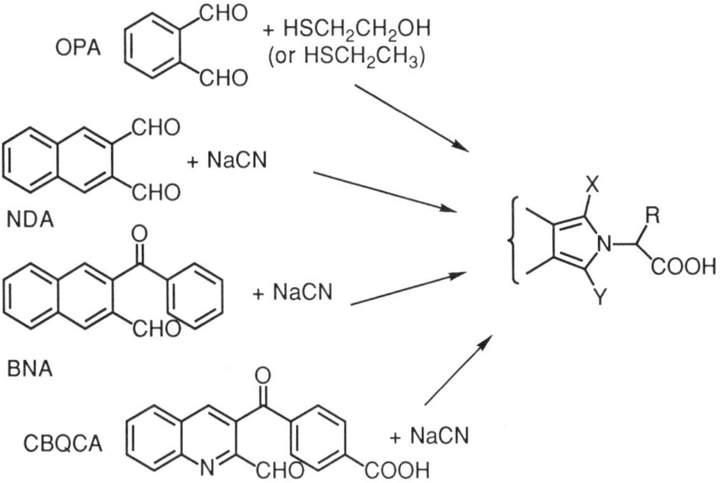

Figure 12.2
Classical and new isoindole-forming derivatizing agents. (OPA = ortho-phthaldialdehyde, NDA = 2,3-naphthalenedialdehyde, BNA = 3-benzoyl-2-naphthaldehyde, CBQCA = 3-(4-carboxybenzoyl)-2-quinolinecarboxaldehyde).

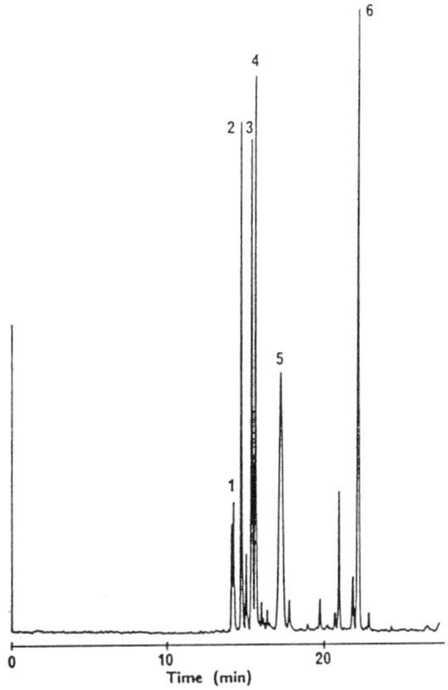

1 = 1-amino-1-deoxyglucose
2 = 1-amino-1-deoxygalactose
3 = 2-amino-2-deoxyglucose
4 = 2-amino-2-deoxygalactose
5 = 6-amino-6-deoxyglucose
6 = D-galactosaminic acid

Figure 12.3
MEKC separation and laser-induced fluorescence detection (λ_{exc} = 442 nm) of CBQCA-derivatized amino sugars +with an anionic surfactant. Capillary : 50 µm x 50/80 cm. Buffer : 50 mM SDS in 20 mM phosphate/borate (pH 9.12). Applied voltage : 16 kV. *(Reprinted with permission [12] © American Chemical Society)*

12.1.2 Dansyl[1] and Dabsyl Derivatives

In a recent review on the electrophoretic separation of proteins, peptides and amino acids [13], it was mentioned that it took several years of intensive research in LC to optimize the complete separation of all amino acids. Nevertheless, the separation of 20 dansylated amino acids has already been demonstrated in MEKC (**Figure 12.4**). Serine and asparagine could be separated by increasing the SDS-concentration to 150 mM but then the resolution between lysine and tyrosine is lost.

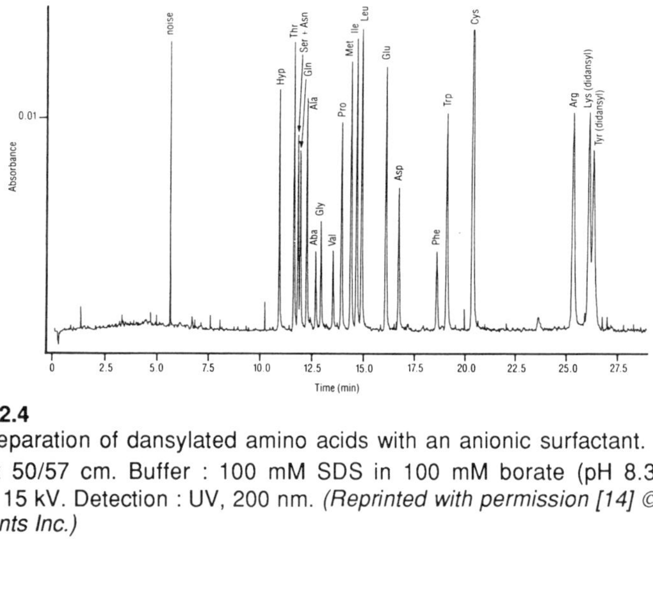

Figure 12.4
MEKC separation of dansylated amino acids with an anionic surfactant. Capillary : 75 µm x 50/57 cm. Buffer : 100 mM SDS in 100 mM borate (pH 8.3). Applied voltage : 15 kV. Detection : UV, 200 nm. *(Reprinted with permission [14] © Beckman Instruments Inc.)*

Bushey and *Jorgenson* described the MEKC analysis of dansylated amines and were able to separate dansylated methylamine and methyl-d_3-amine in a 25 mM SDS/20% MeOH buffer [15,16]. Higher (non-isotopically substituted) linear amines elute close to t_{MC} from C_6 on and require the addition of 30% MeOH, resulting in reversed direction elution mode at a pH-value of 8.0 [16].

The absorbance maximum of dabsyl derivatives overlaps with the blue

[1] *For chiral separations of dansylated D,L-amino acids see also 13.3 and 13.4.3.*

spectral lines of He/Cd and Argon-ion lasers [17] which makes them suitable for thermo-optical detection[1]. In a mixture of 15 dabsyl amino acids, 13 different amino acids were at least partially separated [19] in a neutral buffer containing 50% acetonitrile and 5-10 mM SDS. *Yu* and *Dovichi* expressed doubts whether such a low surfactant concentration would produce micelles but confirmed that the SDS was essential to obtain resolution [19]; however, they contradicted this statement in a subsequent report [20]. The separation quality cannot be compared with that obtained in other systems and the main interest in using dabsyl amino acids is a matter of detectability. A detection limit of 5-50 x 10^{-8} M has been reported [20], representing 40-400 amol of injected amino acid derivative.

12.1.3 Thiocarbamyl and Thiohydantoin Derivatives

Phenylthiohydantoin (PTH)-amino acids[2] are neutral, unless a charge-bearing group is included in the amino acid side chain. *Otsuka* et al. achieved the separation of 22 PTH-amino acids with an anionic surfactant [21], although 4 pairs were poorly resolved (**Figure 12.5**). Those pairs were well resolved with a cationic surfactant (DTAB), although the overall resolution was less satisfactory [21]. Improvement in resolution was also reached by addition of urea to the SDS-containing buffer [22]. When the reference chromatogram (without urea) is compared with the data from [21], some elution order changes can be observed (**Figure 12.6**). Three of the changes involve thiohydantoins that contain a charged side chain (glutamic acid, aspartic acid and histidine) and their migration is only partly determined by their capacity factor [21]. Small changes in $\tilde{k}'$, in this case probably related to a difference in analysis temperature, will affect the migration behavior of neutral compounds more or less proportionately, but will upset the balance between electrophoretic and chromatographic migration of charged compounds, hence the elution order reversals.

The Edman degradation for protein sequencing with phenyl isothiocyanate (PITC) (**Figure 12.7**) is based on the determination of thiohydantoin-amino acids [2]. The small sample requirements in capillary electrophoretic techniques and more sensitive detection can reduce the amount of protein needed for amino acid

---

[1] *Thermo-optical detection [17,18] consists of creating a temperature rise in the capillary with a modulated pump laser (He/Cd or Argon-ion). The resulting thermal lens defocuses the probe laser beam (He/Ne) which is monitored with a small-area photodiode.*

[2] *For the chiral separation of PTH-D,L-amino acids see also 13.4.*

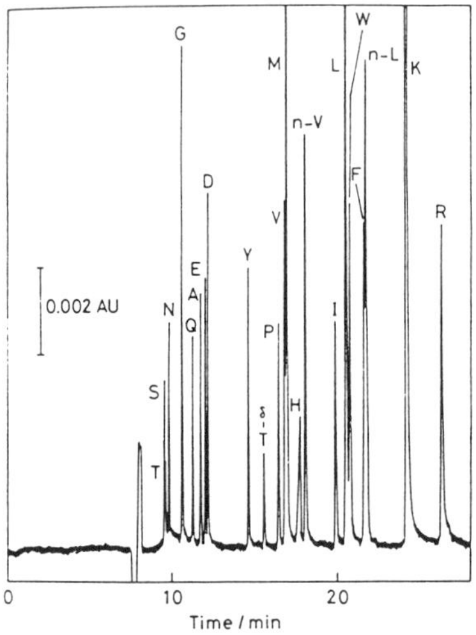

Figure 12.5
MEKC separation of 22 PTH-amino acids with an anionic surfactant. Capillary : 50 µm x 50/65 cm. Buffer : 50 mM SDS in phosphate/borate (pH 7.0). Current : 14 µA (ca. 10 kV, at 35°C). Detection : UV, 260 nm. Amino acids are indicated by their one-letter abbreviations. *(Reprinted with permission [21] © Elsevier Science Publishers B.V.)*

sequence determination [52]. To improve mass sensitivity, other derivatization reagents than PITC have been developed and their use in electrokinetic analysis has been explored.

Derivatization with fluorescein isothiocyanate (FITC) results in negatively charged derivatives. Impressive detection limits have been obtained in the CZE analysis of the fluorescein thiocarbamyl derivatives of amino acids with laser induced fluorescence detection [24,25] but separation has only been demostrated for limited sets of amino acids [4,24,25]. Recently, electropherograms of the fluorescein thiohydantoins (FTH) have been realized in which 15 amino acids could be identified, although not completely separated [23]. Only preliminary results on the MEKC analysis of FTH-amino acids have been described [26].

Dimethylaminoazobenzene thiohydantoins (DABTH) are neutral but can be protonated in acid buffers. Otherwise they are similar to the dabsyl derivatives already discussed and 15 amino acids could be identified, using comparable analytical conditions (40% acetonitrile and 10 mM SDS in a pH 2.5 phosphate buffer) [23].

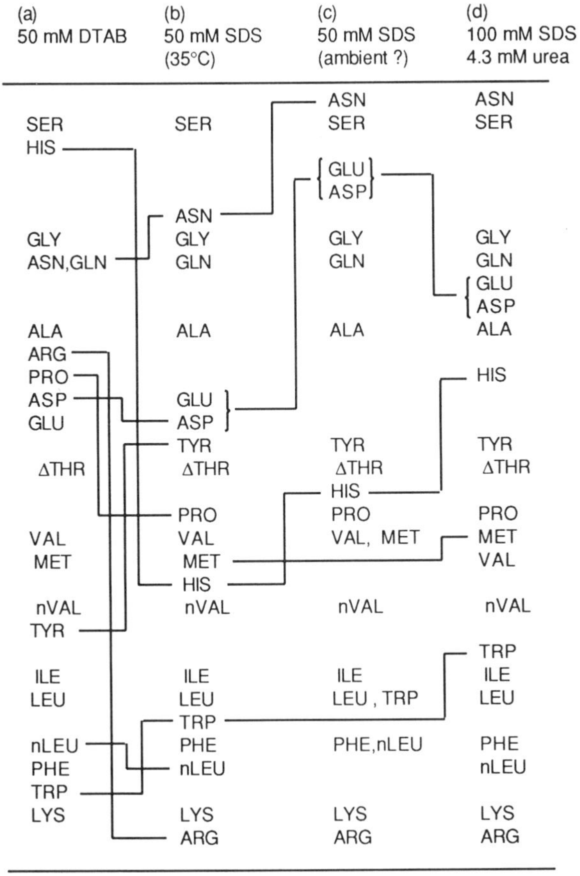

Figure 12.6
Elution order of PTH-amino acids in various media (pH 7.0), compiled from [21] (a,b) and [22] (c,d). Not included are THR, only in [21], and 2-aminoisobutyric acid and 2-aminobutyric acid, only in [22].

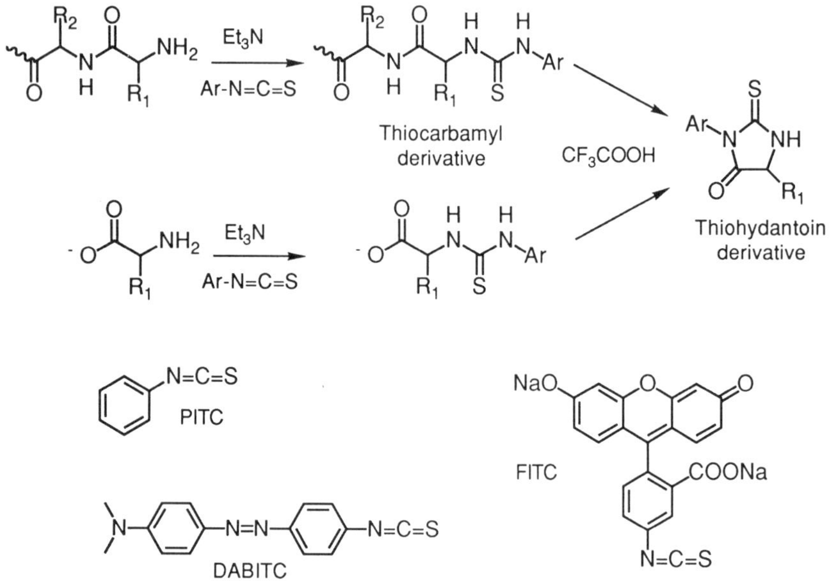

Figure 12.7
N-terminal Edman degradation of proteins, derivatization of amino acids and some isothiocyanate derivatization reagents.

12.1.4 Other Derivatives

Sixteen unidentified amino acid N-2,4-dinitrophenyl-hydrazones (DNPH) have been separated with a 50 mM SDS-buffer [27] but the separation was not studied in detail. MEKC of DNPH derivatives have also been applied to study the Maillard reaction products of glycine and glucose [28].

Fluorescamine derivatives of amines and amino acids are charged and can be separated with CZE [29]. In MEKC (with SDS), amino acid [4] as well as polyamine derivatives [30] showed broadened zones. For the separation of putrescine, cadaverine, spermidine and spermine, the addition of 0.1% ethylene diamine to the buffer resulted in sharper peaks. This method was applied to the determination of polyamines in rat stomach tissue.

NBD (4-chloro-7-nitrobenzo-2-oxa-1,3-diazole) derivatives of aliphatic amines have been used as test substances in various MEKC studies by *Sepaniak* and coworkers [31-38], often in relation to the problem of separating hydrophobic compounds.

12.2 Peptides

The use of surfactants in the classical electrophoresis of proteins is based on the denaturing action through the formation of protein-SDS complexes. These complexes are characterized by a more or less constant binding ratio between surfactant monomers and the number of amino acid residues [39]. In many of the proposed models [40], micellar structures do play a role and several of them are associated with the protein. In a genuine MEKC mechanism, one would expect that micelles do not form superstructures, imposing an upper limit on the peptide size that a single micelle can accommodate. A structure has been proposed [39,41] in which hydrophilic peptide sections are associated with the polar groups at the micelle surface, while hydrophobic sections or side groups penetrate the micelle core (**Figure 12.8**). The upper limit in peptide size at which multiple micelle associations will start to occur cannot yet be defined.

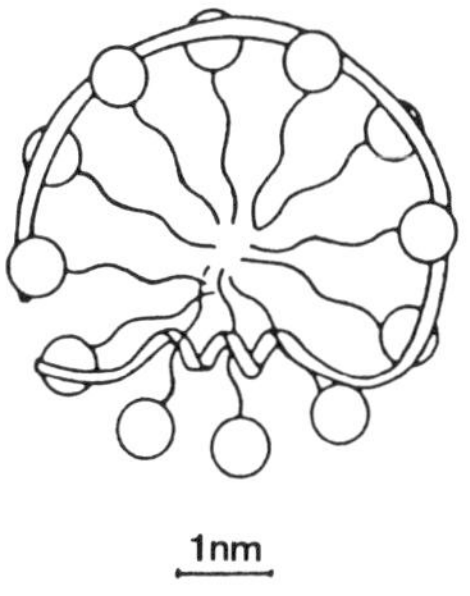

Figure 12.8
Schematic representation of the complex between an SDS micelle and a peptide *(Reprinted with permission [41] © Elsevier Science Publishers B.V.))*

Brownlee and *Compton* showed the separation of 9 oligopeptides (n = 2, 3, 4) using an alkaline buffer containing 200 mM SDS [42]. *Wainright* used SDS to separate several oligopeptides differing in structure only by neutral amino acid residues that could not be separated by CZE alone [43]. On addition of SDS, the elution order between the positively charged angiotensins and the negatively charged enkephalins was reversed. Similarly, from the data on adrenocorticotropic hormone-related fragments by *Van de Goor* et al. [44] it can be inferred that the migration of positively charged peptides is more influenced than that of nearly neutral peptides while negatively charged peptides are almost unaffected. The gain in selectivity, however, is largest for nearly neutral species.

Liu et al. clearly illustrated the improved resolution of cationic peptides (angiotensin analogues) with cationic surfactants (DTAB, HTAB) by comparing the analysis at surfactant concentrations above and below the CMC (**Figure 12.9**) [45]. They also demonstrated separations of OPA-derivatized peptides with SDS.

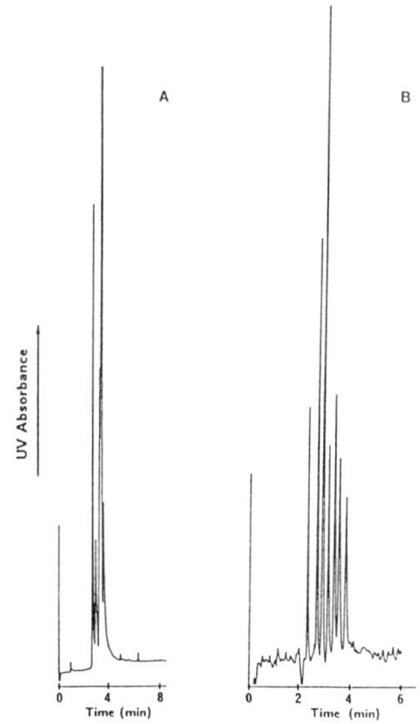

Figure 12.9
Separation of angiotensin analogues with a cationic surfactant (DTAB) at different concentrations. Buffer : 2 mM (A) or 50 mM (B) DTAB in 10 mM phosphate/Tris (pH 7.05). The CMC of DTAB in water is 15 mM. Capillary : 50 μm x 25/45 cm. Applied voltage : -15 kV. Detection : UV, 220 nm. *(Reprinted with permission [45] © Elsevier Science Publishers B.V.)*

Swedberg obtained improved resolution between two heptapeptides after the addition of the non-ionic octyl glucoside [46]. Although *Swedberg* supposed double layer formation at the micelle surface to explain the separation, it can equally well be considered that the slightly more hydrophobic Ile^5-angiotensin III acquires a lower electrophoretic mobility by more efficient association with the neutral micelles than the Val^5-angiotensin III [1].

12.3 Nucleic Acid Related Compounds

12.3.1 Purine and Pyrimidine Bases[2] , Nucleosides

The purine and pyrimidine bases cover a wide range of pKa values and require a pH gradient (from 3.5 to 2.2) to separate several of them as cations in CZE [47] (the mixture did not contain the acidic uracil and thymine). The MEKC separation of the most frequently occurring nucleobases with SDS has been demonstrated [48]. At neutral pH, pyrimidines elute before the purines (in the order U, C, T, G, A [48,49]). In more alkaline buffers (pH 9.0), however, the elution order can change due to ionization of the more acidic species like guanine or uracil [50]. MEKC has been used for separating the normal bases from the alkylated bases resulting from exposure of DNA to epoxy compounds [4]. The ribonucleosides [48,50,51] and deoxyribonucleosides [49,52,53] can be separated under similar conditions (**Figure 12.10**).

12.3.2 Nucleotides and Oligonucleotides

Unlike nucleobases and nucleosides, the nucleotides are strong electrolytes, charged over a wide pH range, and can be analyzed in CZE. When eluted against the electroosmotic flow, monophosphates elute first (lowest electrophoretic mobility), followed by the diphosphates and finally the triphosphates [54]. Without or with reduced electroosmosis and with reversed polarity, the elution order is reversed [55,56]. Simultaneous resolution optimization of mono-, di- and triphosphates within short analysis time remains difficult to achieve [54]. This is one of the reasons why the MEKC analysis of nucleotides has been considered.

Liu et al. compared one anionic (SDS) and two cationic (DTAB and HTAB) surfactants for the analysis of 12 mono-, di- and triphosphates [57]. Using a short column (L_d = 26 cm), they were able to separate the 12 nucleotides with 100 mM

---

[1] *The use of neutral surfactants in relation to MEKC has already been discussed. See section 5.3.*

[2] *For other purine and pyrimidine compounds see also 12.4.2. (Barbiturates), 12.4.5. (Caffeine) and 12.4.6. (Xanthines and uric acid derivatives).*

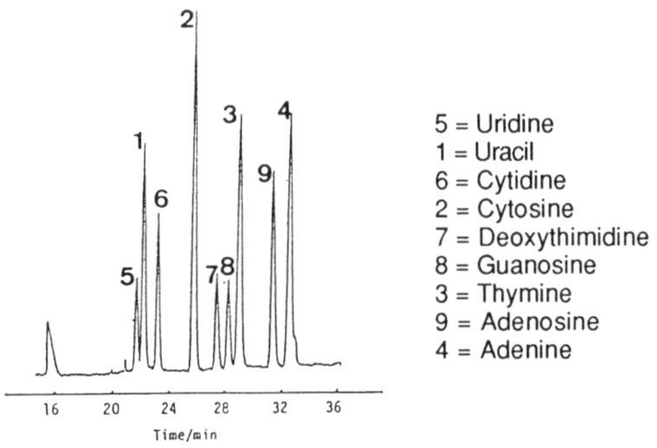

Figure 12.10
MEKC separation of nucleobases and nucleosides. Capillary : 50 µm x 50/65 cm. Buffer : 300 mM SDS in borate/phosphate (pH 7.0). Applied voltage : 5.9 kV. Detection : UV, 210 nm. *(Reprinted with permission [48] © American Chemical Society)*

DTAB at pH 7.0 within 4 minutes. Mixtures of nucleosides, 5'-monophosphates, deoxy-5'-monophosphates together with less common nucleotides and AZT (3'-azido-3'-deoxythymidine), the well-known anti-AIDS drug, have been separated with SDS as the surfactant [58] (**Figure 12.11**). The separation of deoxy-3'-monophosphates with SDS has been applied to enzymatic DNA digests [49].

For high detectability of 2'-deoxy-5'-monophosphates, *Lee* et al. [59] applied 5'-phosphoramidation with ethylenediamine, followed by dansylation, to an enzymatic DNA digest. Using MEKC separation (45 mM SDS, pH 9.0) and laser-induced fluorescence detection, they were able to determine abnormal base residues at the attomole-level (10^{-10} M).

As is the case for peptides, the small dimensions make micelles less suitable for the solubilization of large aggregates. Hence the application to biopolymers is restricted to the smallest class members, in this case the oligonucleotides. In CZE, single base-unit resolution is limited to approximately the decamer [48,55] [1] and addition of a surfactant does not improve this range [48]. On addition of bivalent cations (0.3-5 mM Cu^{2+}, Zn^{2+}, Mg^{2+}) to the SDS-solution however, separation of deoxypolythymidines up to the 18-mer becomes possible [48]. It is presumed that

[1] *For high molecular weight oligonucleotides and DNA or RNA fragments, Capillary Gel Electrophoresis is the capillary electrophoretic technique of choice, allowing resolution up to the 430-mer [60].*

the cations are adsorbed at the surface of the micelle, and in turn, adsorb reversibly the oligonucleotides. As Cu^{2+} ions are more effective than Mg^{2+}, it is probable that the coordination occurs at the base moiety rather than with the phosphate groups. The same methodology has been used for the separation of 18 octamers with different base composition [48].

Some improvement was observed in the MEKC separation of polycytidines on addition of 3 mM spermine but separation beyond the decamer was not demonstrated [55].

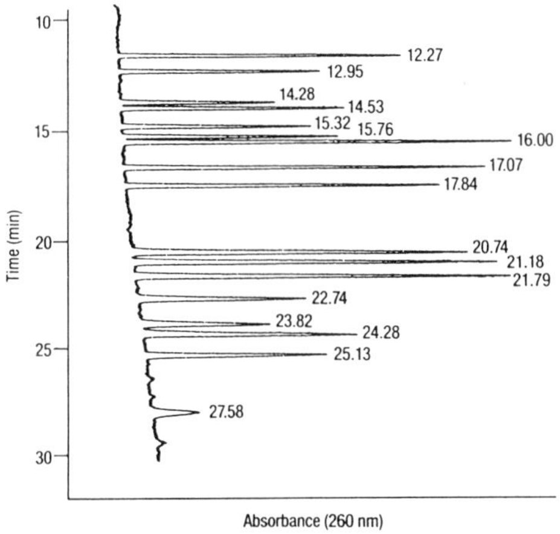

12.27 = Uridine
12.95 = Cytidine
14.28 = Cytosine
14.53 = Thymidine
15.32 = Thymine
15.76 = 2'-Azido-2'deoxycytidine
16.00 = Adenosine
17.07 = 2'-Deoxyguanosine
17.84 = 2'-Deoxyadenosine
20.74 = 5'-GMP
21.18 = 2'-Deoxy-5'-GMP
21.79 = 2'-Deoxy-5'-AMP
22.74 = 2'-Deoxy-5'-CMP
23.82 = 2'-CMP
24.28 = 3'-CMP
25.13 = AZT

Figure 12.11
MEKC separation of nucleosides and nucleotides with an anionic surfactant. Capillary : 75 µm x 60/66 cm. Buffer : 200 mM SDS in 10 mM phosphate buffer (pH 6.86). Applied voltage : 13 kV. Detection : UV, 260 nm.
(Reprinted with permission [58] © Beckman Instruments Inc.)

12.4 Drugs and Related Compounds, Clinical Analyses

12.4.1 Vitamins and some of their Metabolites

Vitamins are a chemically heterogeneous group (**Table 12.1**). Many of the water soluble vitamins can be separated by CZE but as soon as more than one neutral species is present, the use of MEKC is imposed (**Figure 12.12**) [61,62].

Burton et al. separated Vitamin B6 and its metabolites with laser-induced

Table 12.1 MEKC applied to Vitamins

A	Retinol	[66]
B1	Thiamine, Aneurine	[61,62,64-67]
B2	Riboflavin	[27,62,64,66,67]
	Riboflavin phosphate	[61,62]
B3	Nicotinamide	[61,62,64]
B3'	Nicotinic acid, Niacin	[61,62,66,67]
B6	Pyridoxine	[27,61-64,66,67]
	Pyridoxine metabolites	[27,62,63]
B12	Cyanocobalamin	[61,62,64-66]
C	Ascorbic acid	[61,66,67]
E	α-Tocopherol	[66]
H	Biotin	[66]

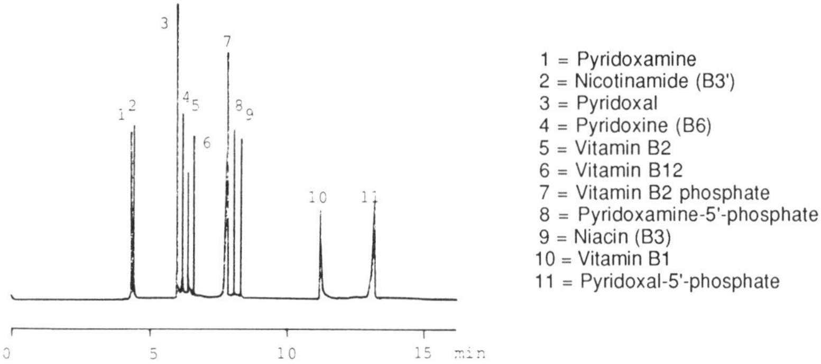

Figure 12.12
Separation of water-soluble vitamins. Capillary : 50 μm x 50/65 cm. Buffer : 50 mM SDS in 20 mM phosphate/borate (pH 9.0). Applied voltage : 25 kV. Detection : UV, 210 nm. *(Reprinted with permission [62] © Elsevier Science Publishers B.V.)*

fluorescence detection [63], but the quantitative determination in human urine extract suffered from interferences. Limits of detection ranged from 150 to 900 pg with an exception for pyridoxic acid which could be determined at the subpicogram level. Laser-induced fluorescence has also been used for the determinatin of Vitamin B2 and its metabolites in blood extracts [27]. *Fujiwara* et al. used UV detection to determine various vitamins in a commercial formulation [61] and *Kobayashi* et al. used diode array detection to identify the components in vitamin tablets [64].

Vitamin mixtures typically consist of molecules with different charge states, leading to irregularities in their retention behavior. The effects of pH and surfactant concentration reflect differences in the interaction with the pseudo-stationary phase [61,62,65,66]. In contrast to nonionic compounds, the retention time ratios of anionic, and especially cationic components were affected by the use of lauroylmethyl taurate instead of SDS or by the addition of anionic ion-pairing reagents [62]. Vitamin B1 and B12 were also selected to study the effect of tetraalkylammonium salts on the retention behavior of ionic and zwitterionic compounds [65].

The problems with hydrophobic compounds have already been mentioned on several occasions. Nevertheless, the simultaneous analysis of water-soluble and two fat-soluble vitamins (A and E) has been demonstrated in CD-MEKC, using γ-cyclodextrin [66].

12.4.2 Barbiturates[1]

Barbiturates have been analyzed as anions with SDS as the surfactant [43,68]. *Wainright* observed a correlation between the elution order and the hydrophobicity in the side chain [43]. This is confirmed by the data presented by *Thormann* et al. [68] (**Table 12.2**). They applied MEKC to compare extraction procedures from spiked blood, using multiwavelength detection to identify

Table 12.2 Relation between MEKC retention order and number of carbons in the side chains of barbiturates.

	pH 8.5, 50 mM SDS [43]	R_1	R_2	pH 7.8, 50 mM SDS [68]	R_1	R_2
1	Barbital	C_2	C_2	Barbital	C_2	C_2
2	Aprobarbital	C_3 (=)	C_3	Allobarbital	C_3 (=)	C_3 (=)
3	Butabarbital	C_4	C_2	Phenobarbital	Phenyl	C_2
4	Amobarbital	C_5	C_2	Butalbital	C_4	C_3 (=)
5	Hexobarbital	C_6 (=, c)	C_1	Thiopental	C_5	C_2
6	Pentobarbital	C_5	C_2	Amobarbital	C_5	C_2
7	Secobarbital	C_5	C_2 (=)	Pentobarbital	C_5	C_2
8	Methohexital	C_6 (≡)	C_2 (=)			

(=) : contains double bond, (≡) : contains triple bond, (c) = cyclic

[1] *For chiral separations of barbiturates, see also 13.3.*

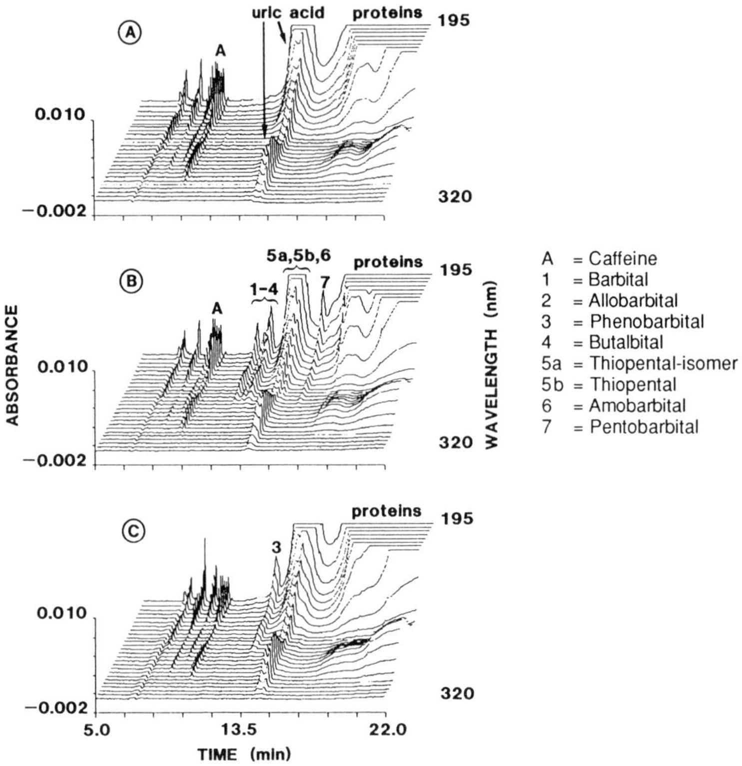

Figure 12.13
MEKC analysis of human serum with multiwavelength detection. (A) Blank serum, (B) Serum spiked with barbiturates, (C) Serum from a patient undergoing phenobarbital pharmacotherapy. Capillary : 75 µm x 70/90 cm. Buffer : 50 mM SDS in borate/phosphate (pH 7.8). Applied volatge : 20 kV. *(Reprinted with permission [68] © Elsevier Science Publishers B.V.)*

barbiturate components in human blood and urine. Direct analysis of low barbiturate levels (µg/ml) in urine was not possible because of the many interferences. In serum, the feasibility of direct analysis was limited to those barbiturates eluting in a useful window, determined by caffeine and uric acid (**Figure 12.13**).

12.4.3 β-Lactam Antibiotics : Penicillins and Cephalosporins

Many penicillins and cephalosporins (**Figure 12.14**) are anionic species and some, like Penicillin G (Benzylpenicillin) [69] or selected cephalosporins [43], can be analyzed with CZE, but complete resolution of complex mixtures cannot always be obtained [70,71]. Because of the side chain structure, some components may have different, including neutral and opposite, charge states. Their precursors, 6-aminopenicillanic and 7-aminocephalosporanic acid, are zwitterions.

The MEKC separation of 7 penicillins and 9 cephalosporins has been studied by *Nishi* et al. [70] (**Figure 12.15**). They concluded that the selectivity is determined by the number of negatively charged groups (carboxylate or sulfonate), lipophilicity and ion-pair formation between the micelles and cationic groups. As the relative importance of these effects will differ for each component, this leads to a strong dependence of the elution order on the surfactant concentration. The importance of ion-pair formation can be reduced by using lauroylmethyl taurate instead of SDS [70] or by the addition of tetraalkylammonium salts [65].

Nakagawa et al. applied MEKC for the analysis of cefpiramide, a cephalosporin, in human plasma with direct injection [72,73]. The presence of SDS in the buffer has several advantages. It promotes release of protein-bound solutes, solubilizes the proteins which results in less column contamination and better reproducibility and finally selectively retards protein migration. The latter effect has also been observed in other plasma analyses [68,71,74]. With antipyrine as an internal standard, linear calibration plots were obtained between 10 and 300 μg/ml, covering usual plasma levels during therapy [73]. Similarly, *Nishi* et al. determined aspoxicillin in human plasma (**Figure 12.16**) with a limit of detection equal to 1.3 μg/ml at a signal-to-noise ratio of 3 [71].

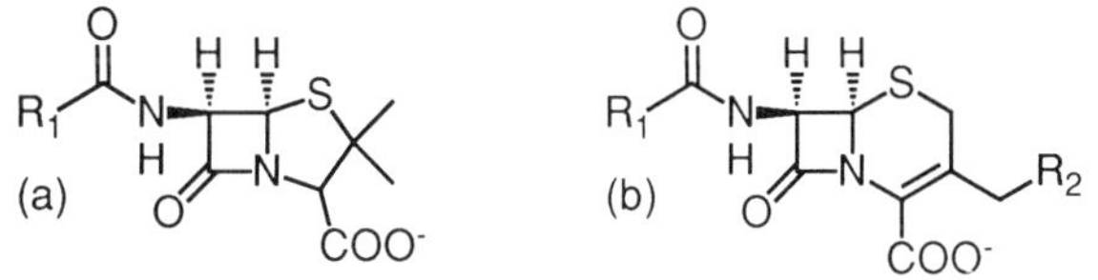

Figure 12.14
β-Lactam antibiotics. (a) Penicillins, (b) Cephalosporins.

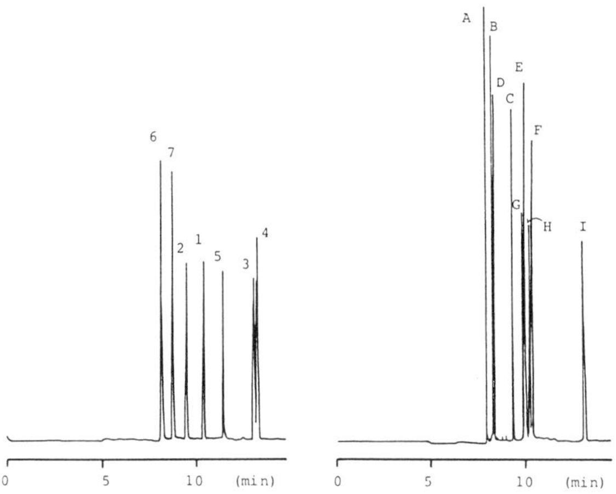

Figure 12.15

MEKC separation of penicillins (left) and cephalosporins (right). Capillary : 50 µm x 50/65 cm. Buffer : 150 mM SDS in 20 mM phosphate/borate (pH 9.0). Applied voltage : 20 kV. Detection : UV, 210 nm. *(Reprinted with permission [70] © Elsevier Science Publishers B.V.)*

1 = Benzylpenicillin	A = Ceftazidime
2 = Ampicillin	B = Cefotaxime
3 = Carbenicillin	C = Cefoperazone
4 = Sulbenicillin	D = Cefmenoxime
5 = Piperacillin	E = Cefpiramide
6 = Aspoxicillin	F = Ceftriaxone
7 = Amoxicillin	G = Cefpimizole
	H = Cefminox
	I = C-TA

12.4.4 Steroids

Steroids are too lipophilic to be analyzed under normal conditions, i.e. MEKC with SDS, as they migrate with the micelles [75,76]. The methods that can be used for the MEKC analysis of hydrophobic compounds have already been discussed. All of them have been applied in the analysis of steroid compounds (**Figure 12.17**).

Nishi et al. used bile salt surfactants for the analysis of corticosteroids [75] and found optimal resolution with sodium cholate or its taurine-conjugate. The migration

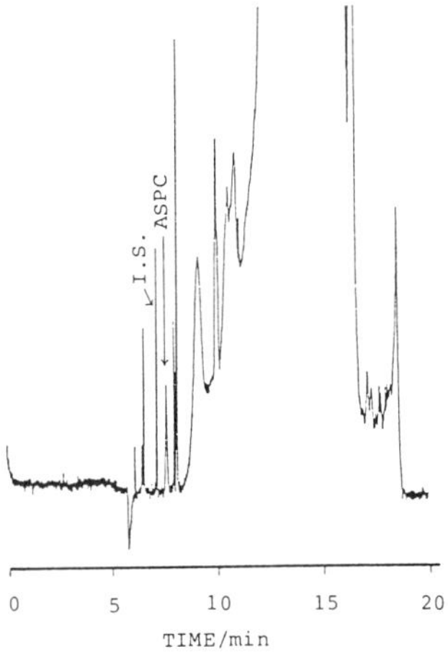

Figure 12.16
MEKC determination of aspoxicillin (ASPC) in human blood plasma. I.S. is the internal standard (acetaminophen). Capillary : 50 µm x 50/65 cm. Buffer : 50 mM SDS in 20 mM phosphate/borate (pH 8.5). Applied voltage : 20 kV. Detection : UV, 210 nm. *(Reprinted with permission [71] © Elsevier Science Publishers B.V.)*

order was well correlated with hydrophobicity. The analysis was applied to the determination of fluocinonide in a corticosteroid cream with a benzothiazepin as internal standard. The unusual high variability in the peak area ratio (9.3%) was hypothetically related to interference from a nonionic surfactant in the formulation.

A very similar set of corticosteroids was used for the separation with SDS and with the addition of 6 M urea to the buffer [22]. The presence of urea increases the solubility of the corticosteroids and enlarges the elution range, resulting in a dramatically improved resolution.

A third possibility, CD-MEKC with SDS, was demonstrated by *Nishi* and *Matsuo* [76]. As urea is also added (4 M) to solubilize the cyclodextrins [77] some overlap with the above mentioned method must be considered. With β-cyclodextrins, capacity factors were reduced but with γ-cyclodextrin, whose larger cavity allowed more effective inclusion, the elution order changed.

Testosterone esters are more hydrophobic than the corticosteroids. *Vindevogel* and *Sandra* reported the separation of four esters from a commercial formulation with SDS and an organic modifier, 35% v/v of acetonitrile [78].

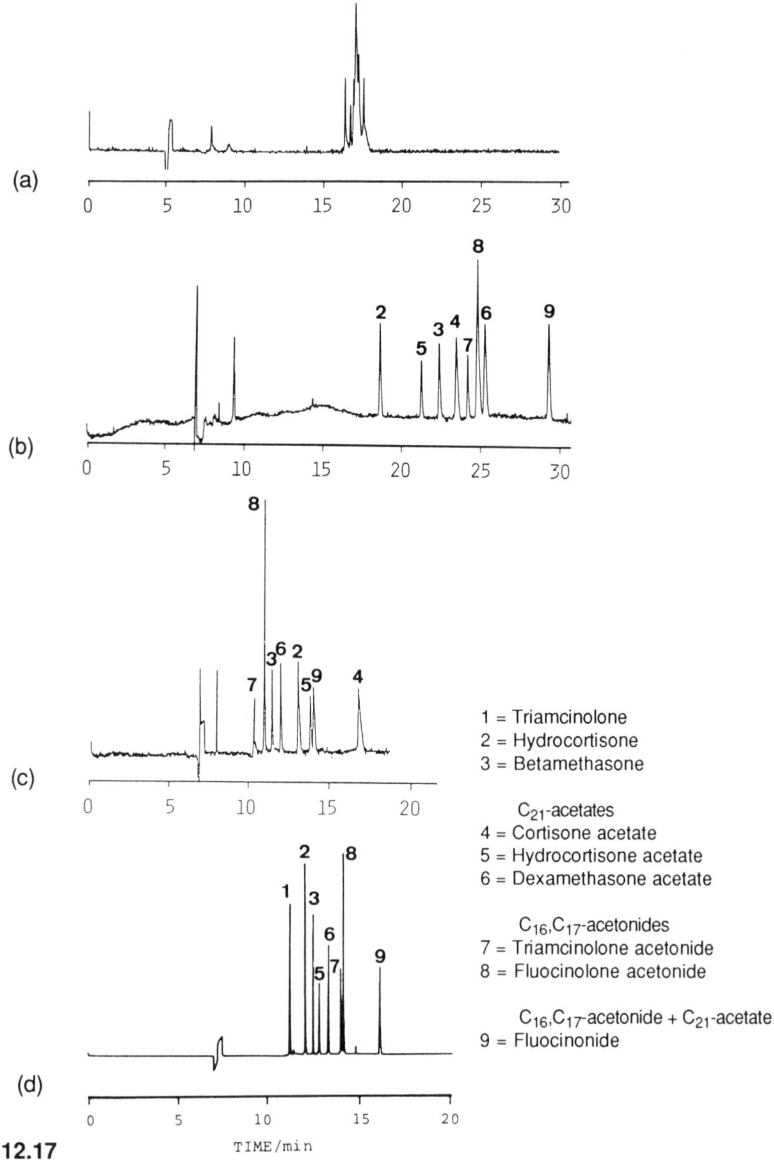

Figure 12.17

MEKC analysis of corticosteroids. Capillary : 50 μm x 50/65 cm. Buffer : 20 mM phosphate/borate (pH 9.0).with : (a) 50 mM SDS, (b) 50 mM SDS, 6 M urea, (c) 50 mM SDS, 4 M urea, 15 mM γ-cyclodextrin. (d) 100 mM sodium cholate. Applied voltage : 20 kV. Detection : UV, 210 or 220 nm. *(Reprinted with permission, a,b-[22] and d-[75] © Elsevier Science Publishers B.V., c-[76] © Marcel Dekker Inc.)*

12.4.5 Analgesics, Anesthetics, Antipyretic and Non-Steroidal Anti-Inflammatory[1] Drugs

The pharmaceuticals discussed in this section are heterogeneous, in a chemical sense as well as in their therapeutic action although some of them are often encountered together e.g. in pain-relieving or cold medicines. Many of them are ionic and can be separated by electrophoresis (e.g. [43]); however, determination of the composition of a particular mixture may often require MEKC.

Fujiwara and Honda determined the constituents in commercial antipyretic analgesic tablets : caffeine, p-acetamidophenol (acetaminophen), o-ethoxybenzamide (ethenzamide) and ethyl p-aminobenzoate [79]. Aspirin and salicylamide, although not present in the tablet, were included in the reference mixture used for separation optimization. The influence of SDS concentration and pH can be related to chemical features of the solutes like the presence of phenolic or basic groups. Similar analyses involving aspirin and related compounds have also been reported [67].

Nishi et al. studied the effect of the surfactant struture on the separation of 14 ingredients from cold medicines [80,81]. With SDS, the complete resolution was difficult as the most hydrophobic and basic components (chlorpheniramine, tipepidine, dibucaine, triprolidine) all coeluted with the micelles. Most complete separation is obtained with bile salts and the different selectivities of cholate and deoxycholate can be used to resolve remaining coelutions (**Figure 12.18**). Four of these compounds (caffeine, acetaminophen, ethenzamide and tipepidine) were quantitatively determined in Novapon granules, using p-hydroxybenzoate as internal standard [81].

Weinberger and *Albin* separated 6 anti-inflammatory carboxylates (naproxen, sulindac, diflunisal, ibuprofen, indomethacin and tolmetin) in a SDS buffer [82]. Although separation could also be obtained with CZE, MEKC was applied to study problems related to linear dynamic range and solute concentration dependent peak broadening.

[1] *For steroidal anti-inflammatory drugs see also 12.4.4.*

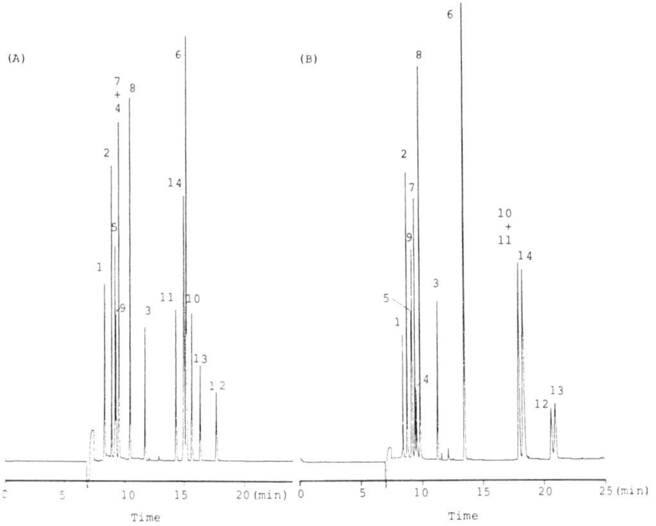

Figure 12.18
MEKC separation of cold medicine ingredients. Capillary : 50 μm x 50/65 cm. Buffer : phosphate/borate buffer (pH 9.0) with 100 mM sodium cholate (A) or 50 mM sodium deoxycholate (B). For peak identification see **Figure 12.19**. *(Reprinted with permission [80] © Elsevier Science Publishers B.V.)*

12.4.6 Various Other Drugs, Metabolites and Physiologically Relevant Compounds

Diltiazem[1], a benzothiazepin vasodilator, is hydrophobic and has been separated from related compounds with the aid of bile salts (SDC,SC) [75]. With MEKC, diltiazem was quantified in tablets and impurities were determined in the range 0.3-0.5%.

Ewing and coworkers have studied the analysis of neurochemicals with the ultimate aim of their determination in individual neurons. The MEKC separation of catechol, 4-methyl- and 4-ethylcatechol and norepinephrine was used to

[1] *For the chiral resolution, see also 13.4.3.*

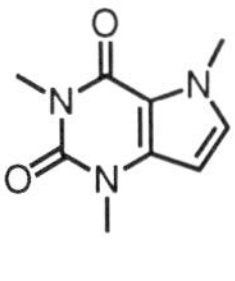

1 = Caffeine

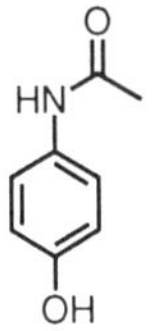

2 = Acetaminophen

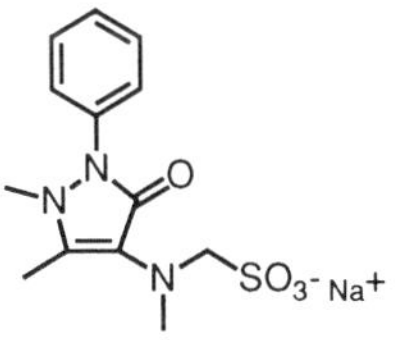

3 = Sulpyrin

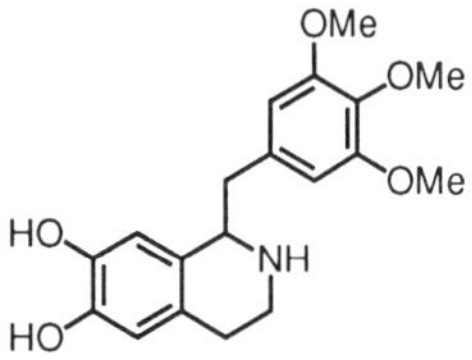

4 = Trimethoquinol (Tretoquinol, as hydrochloride)

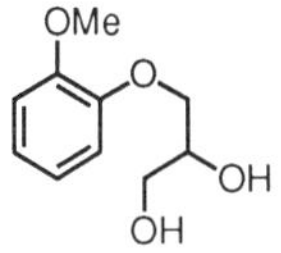

5 = Guaifenesin

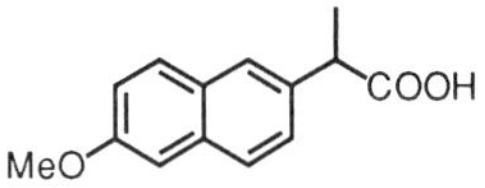

6 = Naproxen

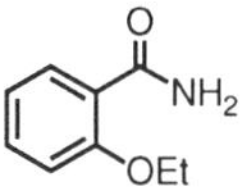

7 = Ethenzamide

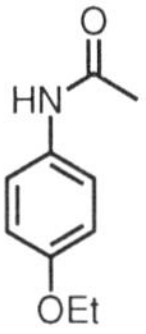

8 = Phenacetin

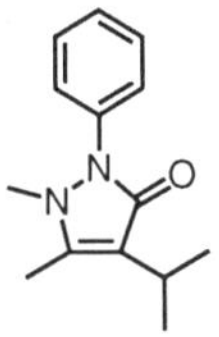

9 = Isopropylantipyrine (Propyphenazone)

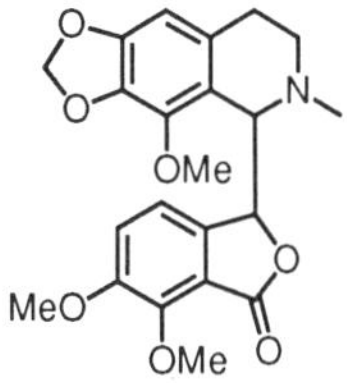

10 = Noscapine

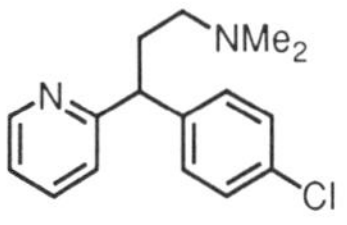

11 = Chlorpheniramine (as maleate)

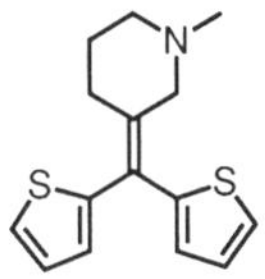

12 = Tipepidine (as hibenzate)

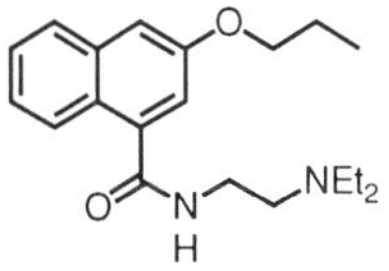

13 = Dibucaine (as hydrochloride)

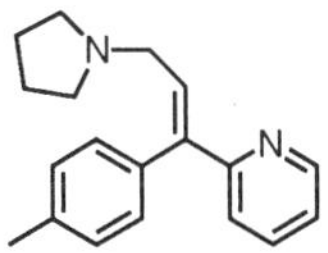

14 = Triprolidine (as hydrochloride)

Figure 12.19
Chemical structures of the pharmaceutical compounds in **Figure 12.18**.

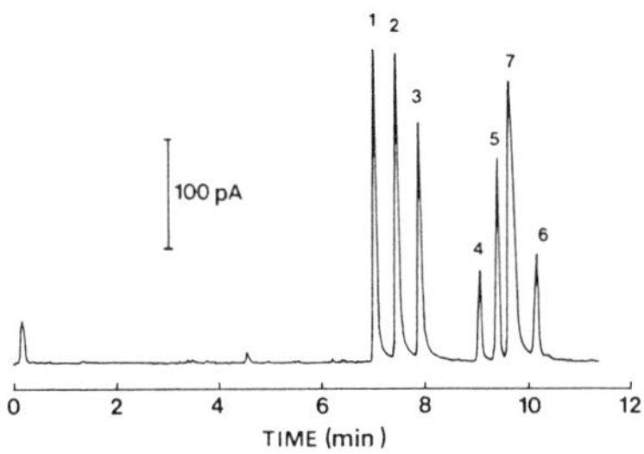

Figure 12.20
MEKC separation and electrochemical detection of catecholamines.
Capillary : 26 µm x 64 cm. Buffer : 10 mM SDS in phosphate/borate (pH 7.0). Applied voltage : 20 kV. *(Reprinted with permission [84] © Elsevier Science Publishers B.V.)*

demonstrate the applicability of electrochemical detection with micellar solutions [83]. The role of borate complexation [84,85] (**Figure 12.20**) and the use of mixed micelles [86] for the separation of mixtures of nonionic and ionic solutes have been studied.

The separation of illicit drugs has been reported by *Weinburger* and *Lurie* [87]. The addition of 15% acetonitrile to the SDS-containing buffer allowed equal distribution of peaks over the available elution range. The migration times of 35 forensically important substances were determined. In the same buffer, the separation of heroin and cocain from their impurities, degradation products and adulterants was demonstrated.

Uroporphyrins are anionic and can be separated by CZE in a methanol containing buffer, but loss of resolution and peak broadening occurs after extended capillary usage [88]. This is ascribed to solubility and wall adsorption problems with some of the less caboxylated porphyrins (mesoporphyrin and coproporphyrin I). Therefore, *Weinberger* et al. added SDS to the buffer [88]. With the exception of the not naturally occurring mesoporphyrin, the separation in the micellar medium was essentially controlled by the electrophoretic mobility in the aqueous phase. With fluorescence detection, the detection limit in spiked urine was estimated at 100-200 pmol/ml. Elevated uro- and heptaporphyrin levels, expected in the urine of patients suffering from *porphyria cutanea tarda*, were observed.

Atamna et al. reported the separation of methyl-substituted uric acids and xanthines [89], caffeine metabolites present in human plasma and urine. Seven methyl-substituted xanthines were completely separated with 15 mM SDS at pH 7.0. For six methyl-substituted uric acids however, the resolution of 1 pair was lost in MEKC although it was obtained in CZE at the same pH. The limited sensitivity of UV detection prevented analysis in biological fluids.

Miyake et al. determined creatinine and uric acid in plasma and urine, with antipyrine as internal standard [74]. The influence of SDS on plasma proteins during analysis with direct injection has already been mentioned. In urine, interference from other endogeneous compounds required adaptation of the analytical conditions. With UV detection clinically relevant concentrations (2.5-10 μg/ml for creatinine and 25-100 μg/ml uric acid in blood, 0.3-1.2 mg/ml for creatinine and 100-400 μg/ml uric acid in urine) could be determined and were compared with data obtained with conventional methods.

12.5 Natural Products

MEKC analysis with SDS as the surfactant has been applied to several plant extracts. Nine flavonol-3-O-glycosides were separated by *Pietta* et al. and six of them could be identified in a *Ginkgo biloba* extract [90]. *Vindevogel* et al. separated the main hop bitter acids, 3 humulones and 3 lupulones, in hop (*Humulus lupulus)* extracts [91,92]. The limited availability of pure reference compounds required the recording of on-line UV spectra for peak identification [91]. Similarly, the cis- and trans-iso-α-acids, present in beer and pre-isomerized hop extracts, were determined [93]. UV detection was not sensitive enough for direct quantification of the iso-α-acids in beer and preliminary extraction was necessary [94].

Aflatoxins are rather hydrophobic and require the addition of organic modifiers like acetonitrile or isopropanol [95]. Reduced analysis time is obtained when SDS is replaced by sodium deoxycholate [38] (**Figure 12.21**).

12.6 Organic Compounds and Technical Products

MEKC was primarily designed for the electrophoretic separation of neutral molecules but, somewhat contradictorily, the number of separations of nonionic compounds is limited. Most common detection methods require the presence of a

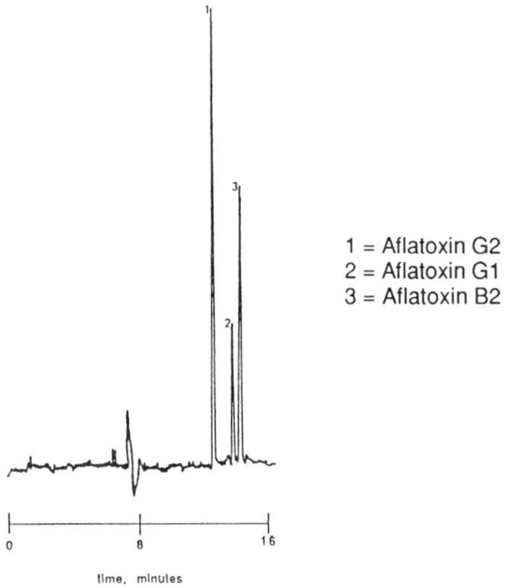

Figure 12.21 :
MEKC separation with laser-induced fluorescence detection (λ_{exc} = 325 nm) of aflatoxins. Buffer : 50 mM SDC in phosphate/borate. (*Reprinted with permission [38] © Elsevier Science Publishers B.V.)*

fluorophore or chromophore, while on the other hand MEKC performs best with rather hydrophilic compounds. Those two conditions are rarely realized simultaneously and seem to be limited to phenols, anilines and nitro compounds [96]. Other compounds often require modifications of the most elementary form of MEKC, i.e. a simple aqueous buffer with an ionic surfactant.

All isomeric chlorinated phenols have been separated by *Otsuka* et al. [97] (**Figure 12.22**). In addition, the MEKC separations of mixtures that also contain substituted cresols [98,99] and nitrophenols [99] have been reported. As can be expected, the migration times of the higher chlorinated phenols are more pH sensitive in the pH range 6.0-9.0 than the less acidic phenol or monochlorophenols [97]. In fact, the dependence of charge on pH can be exploited for the CZE separation of subsets of the chlorophenols [100].

The separation of the lower alkyl parabens (disinfectants, alkyl p-hydroxybenzoates) is easily obtained for the methyl, ethyl, propyl, butyl esters [101] but a more complete separation, including isopropyl-, isobutyl- and isopentyl-esters has been reported with bile salts (sodium deoxycholate) [75].

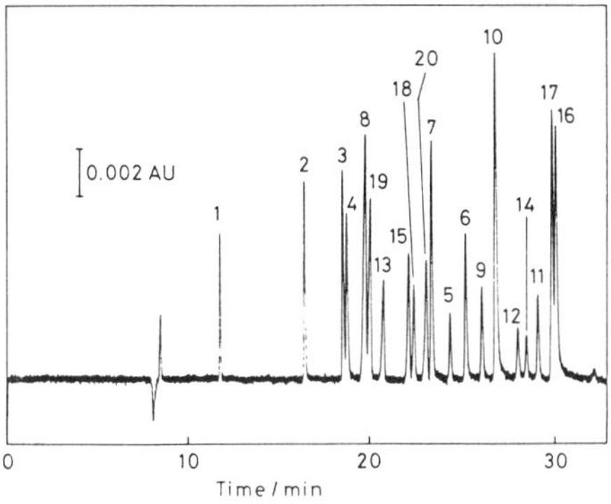

1 = phenol, 2 = 2-Cl-
3 = 2-Cl-, 4 = 3-Cl-
5 = 2,3-Cl_2, 6 = 2,4-Cl_2-
7 = 2,5-Cl_2-, 8 = 2,6-Cl_2-
9 = 3,4-Cl_2-, 10 = 3,5-Cl_2-
11 = 2,3,4-Cl_3-, 12 = 2,3,5-Cl_3-
13 = 2,3,6-Cl_3-, 14 = 2,4,5-Cl_3-
15 = 2,4,6-Cl_3-, 16 = 3,4,5-Cl_3-
17 = 2,3,4,5-Cl_4-, 18 = 2,3,4,6-Cl_4-
19 = 2,3,5,6-Cl_4-, 20 = Cl_5-

Figure 12.22
MEKC separation of all the isomeric chlorinated phenols. Capillary : 50 µm x 50/65 cm. Buffer : 70 mM SDS in phosphate/borate (pH 7.0). Current : 17 µA (ca.10 kV, at 35°C). Detection : UV, 220 nm. *(Reprinted with permission [97] © Elsevier Science Publishers B.V.)*

Several nitrotoluenes, nitronaphthalenes and other nitro-compounds, present in gunpowder and explosives, were separated [102]. By determining these constituents, MEKC can be used to study gunpowder aging or to identify plastic explosives, like Semtex, and gunpowder in spent ammunition cases (**Figure 12.23**).

The separation of benzene and benzaldehyde in the ASTM test mixture LC-79-2, also containing benzyl alcohol, acetophenone, methyl benzoate and dimethyl terephthalate, seems difficult. Resolution has been obtained by adding a nonionic cosurfactant to the SDS phase [103,104] although separation has been obtained with SDS alone [105].

As hydrophobicity increases, the use of bile salts or dedicated additives must be considered. *Cole* et al. used bile salts in combination with 10-20% of methanol to separate polyaromatic hydrocarbons (anthracene, pyrene, benzo[a]pyrene) [38]. *Terabe* et al. separated the 12 chlorinated benzene congeners, trichlorobiphenyl isomers and tetrachlorodibenzo-p-dioxin isomers and polyaromatic hydrocarbons with cyclodextrin added to the buffer (CD-MEKC) [106]. *Terabe* et al. separated polyaromatic hydrocarbons, adding urea to SDS solutions [22]. Urea was also used to promote the separation of benzothiazole sulfenamides (rubber product accelerators) [107].

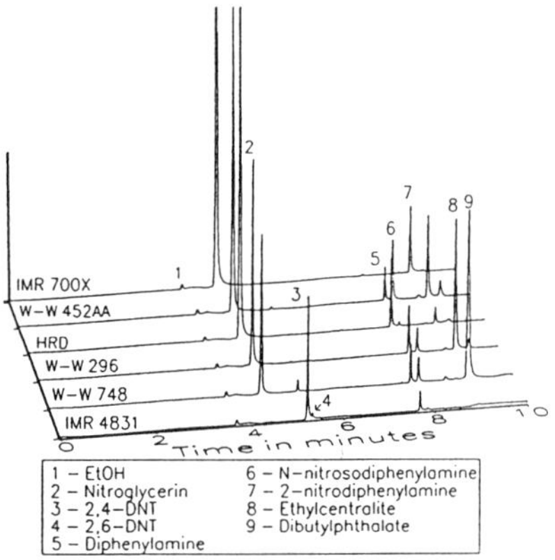

Figure 12.23
Comparative MEKC analysis of several explosive powder extracts. Capillary : 100 µm x 62/67 cm. Buffer : 25 mM SDS in 2.5 mM borate. Applied voltage : 20 kV. Detection : UV, 200 nm. (*Reprinted with permission [102] © American Chemical Society)*

12.7 Metal Chelates

Anionic 4-(2-pyridylazo)resorcinolato complexes of transition metals like Co(III), Cr(III), Ni(II) and Fe(III) can be separated and detected, provided that the chelate is stable [108]. From elution order differences between CZE and MEKC, selective retardation of the chelates by the micellar phase was concluded. Chelation allows detection of metal ions with common UV detectors (the detection limit is estimated at 2.3 x 10^{-6} M) and the applicability to trace metal analysis has been suggested.

Neutral acetylacetonato complexes of Cr(III), Co(III) Rh(III) and Pt(II) or Pd(II) have also been subjected to MEKC [109]. From the observed linearity between $\tilde{k}'$ and SDS-concentration, effective micellization of the metal complexes is concluded. Capacity factors could be predicted from the partition coefficient in the dodecane/water system.

References

[1] W.G.Kuhr, Anal.Chem., **62** (1990) 403R.
Capillary electrophoresis.
[2] K.Blau, G.S.King (eds), "Handbook of Derivatives for Chromatography", Heyden & Son Ltd, London, 1977.
[3] S.L.Pentoney, X.Huang, D.S.Burgi, R.N.Zare, Anal.Chem., **60** (1988) 2625.
On-line connector for microcolumns : Application to the on-column o-phthaldialdehyde derivatization of amino acids separated by capillary zone electrophoresis.
[4] M.Albin, R.Weinberger, E.Sapp, S.Moring, Anal.Chem., **63** (1991) 417.
Fluorescence detection in capillary electrophoresis : evaluation of derivatizing reagents and techniques.
[5] J.Liu, K.A.Cobb, M.Novotny, J.Chromatogr., **468** (1988) 55.
Separation of pre-column ortho-phthalaldehyde-derivatized amino acids by capillary zone electrophoresis with normal and micellar solutions in the presence of organic modifiers.
[6] P.de Montigny, J.F. Stobaugh, R.S.Givens, R.G.Carlson, K.Srinivasachar, L.A.Sternson, T.Higuchi, Anal.Chem., **59** (1987) 1096.
Naphthalene-2,3-dicarboxaldehyde/cyanide ion : a rationally designed fluorogenic reagent for primary amines.
[7] B.Nickerson, J.W.Jorgenson, J.High Resolut.Chromatogr. Chromatogr. Commun., **11** (1988) 533.
High speed capillary zone electrophoresis with laser induced fluorescence detection.
[8] R.T.Kennedy, M.D.Oates, B.R.Cooper, B.Nickerson, J.W.Jorgenson, Science, **246** (1989) 57.
Microcolumn separations and the analysis of single cells.
[9] S.C.Beale, Y.-Z.Hsieh, J.C.Savage, D.Wiesler, M.Novotny, Talanta, **36** (1989) 321.
3-Benzoyl-2-quinolinecarboxaldehyde : a novel fluorogenic reagent for the high-sensitivity chromatographic analysis of primary amines.
[10] Y.-Z.Hsieh, S.C.Beale, D.Wiesler, M.Novotny, J.Microcol.Sep., **1** (1989) 96.
3-Benzoyl-2-naphthaldehyde, a new fluorogenic reagent for microcolumn liquid chromatography-laser-induced fluorescence detection of amino acids.
[11] J.Liu, Y.-Z.Hsieh, D.Wiesler, M.Novotny, Anal.Chem., **63** (1991) 408.
Design of 3-(4-carboxybenzoyl)-2-quinolinecarboxaldehyde as a reagent for ultrasensitive determination of primary amines by capillary electrophoresis using laser fluorescence detection.
[12] J.Liu, O.Shirota, M.Novotny, Anal.Chem., **63** (1991) 413.
Capillary electrophoresis of amino sugars with laser-induced fluorescence detection.
[13] M.V.Novotny, K.A.Cobb, J.Liu, Electrophoresis (Weinheim), **11** (1990) 735.
Recent advances in the capillary electrophoresis of proteins, peptides and amino acids.
[14] Y.Miyashita, S.Terabe, Applications Data DS-767, Beckman Instruments Inc., 1990.
Separation of dansyl DL-amino acids by micellar electrokinetic capillary chromatography with and without cyclodextrins using P/ACE™ System 2000.
[15] M.M.Bushey, J.W.Jorgenson, Anal.Chem., **61** (1989) 491.
Separation of dansylated methylamine and dansylated methyl-d_3-amine by micellar electrokinetic capillary chromatography with methanol-modified mobile phase.
[16] M.M.Bushey, J.W.Jorgenson, J.Microcol.Sep., **1** (1989) 125.
Effects of methanol-modified mobile phase on the separation of isotopically substituted compounds by micellar electrokinetic capillary chromatography.
[17] M.Yu, N.J.Dovichi, Appl.Spectrosc., **43** (1989) 196.
Attomole amino acid determination : capillary zone electrophoresis with laser-based thermo-optical detection.
[18] N.J.Dovichi, Chem.Br., **24** (1988) 895.
Pushing back the limits of detection.
[19] M.Yu, N.J.Dovichi, Mikrochim.Acta, 1988, III, 27-40
Sub-femtomole determination of DABSYL-amino acids with capillary zone electrophoresis separation and laser-induced thermooptical absorbance detection.
[20] M.Yu, N.J.Dovichi, Anal.Chem., **61** (1989) 37.
Attomole amino acid determination by capillary zone electrophoresis with thermooptical absorbance detection.
[21] K.Otsuka, S.Terabe, T.Ando, J.Chromatogr., **332** (1985) 219.
Electrokinetic chromatography with micellar solutions. Separation of phenylthiohydantoin-amino acids.

[22] S.Terabe, Y.Ishihama, H.Nishi, T.Fukuyama, K.Otsuka, J.Chromatogr, **545** (1991) 359.
Effect of urea addition in micellar electrokinetic chromatography.

[23] K.C.Waldron, S.Wu, C.W.Earle, H.R.Harke, N.J.Dovichi, Electrophoresis (Weinheim), **11** (1990) 777.
Capillary zone electrophoresis separation and laser-based detection of both fluorescein thiohydantoin and dimethylaminoazobenzene thiohydantoin derivatives of amino acids.

[24] Y.F.Cheng, N.J.Dovichi, Science, **242** (1988) 562.
Subattomole amino acid analysis by capillary zone electrophoresis and laser-induced fluorescence.

[25] S.Wu, N.J.Dovichi, J.Chromatogr., **480** (1989) 141.
High-sensitivity fluorescence detector for fluorescein isothiocyanate derivatives of amino acids separated by capillary zone electrophoresis.

[26] B.Nickerson, J.W.Jorgenson, Poster presented at the 2nd International Symposium on High Performance Capillary Electrophoresis, San Francisco, January 29-31,1990 (P-216).
Separation of fluorescein thiohydantoin amino acids by micellar electrokinetic chromatography.

[27] D.F.Swaile, D.E.Burton, A.T.Balchunas, M.J.Sepaniak, J.Chromatogr.Sci., **26** (1988) 406.
Pharmaceutical analysis using micellar electrokinetic capillary chromatography.

[28] Z.Deyl, I.Miksik, R.Struzinski, J.Chromatogr., **516** (1990) 287.
Separation and partial characterization of Maillard reaction products by capillary electrophoresis.

[29] J.W.Jorgenson, K.D.Lukacs, Anal.Chem., **53** (1981) 1298.
Zone electrophoresis in open-tubular glass capillaries.

[30] T.Tsuda, Y.Kobayashi, A.Hori, T.Matsumoto, O.Suzuki, J.Microcol.Sep., **2** (1990) 21.
Separation of polyamines in rat tissues by capillary electrophoresis.

[31] M.J.Sepaniak, R.O.Cole, Anal.Chem., **59** (1987) 472.
Column efficiency in micellar electrokinetic capillary chromatography.

[32] A.T.Balchunas, M.J.Sepaniak, Anal.Chem., **59** (1987) 1466.
Extension of elution range in micellar electrokinetic capillary chromatography.

[33] A.T.Balchunas, M.J.Sepaniak, Anal.Chem., **60** (1988) 617.
Gradient elution for micellar electrokinetic capillary chromatography.

[34] M.J.Sepaniak, D.F.Swaile, A.C.Powell, J.Chromatogr., **480** (1989) 185.
Instrumental developments in micellar electrokinetic capillary chromatography.

[35] M.J.Sepaniak, D.F.Swaile, A.C.Powell, R.O.Cole, J.High Resolut. Chromatogr., **13** (1990) 679.
Capillary electrokinetic separations : influence of mobile phase composition on performance.

[36] R.O.Cole, M.J.Sepaniak, W.L.Hinze, J.High Resolut.Chromatogr., **13** (1990) 579.
Optimization of binaphthyl enantiomer separations by capillary zone electrophoresis using mobile phases containing bile salts and organic solvent.

[37] A.C.Powell, M.J.Sepaniak, J.Microcol.Sep., **2** (1990) 278.
Development of a model for predicting retention times in solvent-gradient micellar electrokinetic capillary chromatography.

[38] R.O.Cole, M.J.Sepaniak, W.L.Hinze, J.Gorse, K.Oldiges, J.Chromatogr., **557** (1991) 113.
Bile salt surfactants in micellar electrokinetic chromatography. Application to hydrophobic molecule separations.

[39] P.Lundahl, E.Greijer, M.Sandberg, S.Cardell, K.-O. Eriksson, Biochim. Biophys.Acta, **873** (1986) 20.
A model for ionic and hydrophobic interactions and hydrogen-bonding in sodium dodecyl sulfate-protein complexes.

[40] E.Mascher, P.Lundahl, J.Chromatogr., **476** (1989) 147.
Sodium dodecyl sulfate-protein complexes. Changes in size or shape below the critical micelle concentration, as monitored by high-performance agarose gel chromatogrpahy.

[41] P.Lundahl, E.Mascher, K.Kameyama, T.Takagi, J.Chromatogr., **518** (1990) 111.
Water-soluble proteins do not bind octyl glucoside as judged by molecular sieve chromatographic techniques.

[42] R.G.Brownlee, S.W.Compton, Am.Biotechn.Lab., **6**(7) (1988) 10.
Automated instrumentation for analytical capillary electrophoresis.

[43] A.Wainright, J.Microcol.Sep., **2** (1990) 166.
Capillary electrophoresis applied to the analysis of pharmaceuticals.

[44] T.A.A.M.Van de Goor, P.S.L.Janssen, J.W.Van Nispen, M.J.M.Van Zeeland, F.M.Everaerts, J.Chromatogr., **545** (1991) 379.
Capillary electrophoresis of peptides. Analysis of adrenocorticotropic hormone-related fragments.

[45] J.Liu, K.A.Cobb, M.Novotny, J.Chromatogr., **519** (1990) 189.

Capillary electrophoretic separations of peptides using micelle-forming compounds and cyclodextrins as additives.
[46] S.A.Swedberg, J.Chromatogr., **503** (1990) 449.
Use of non-ionic and zwitterionic surfactants to enhance selectivity in high-performance capillary electrophoresis. An apparent micellar electrokinetic capillary chromatographic mechanism.
[47] V.Šustáček, F.Foret, P.Boček, J.Chromatogr., **480** (1989) 271.
Simple method for generation of a dynamic pH gradient in capillary zone electrophoresis.
[48] A.S.Cohen, S.Terabe, J.A.Smith, B.L.Karger, Anal.Chem., **59** (1987) 1021.
High-performance capillary electrophoretic separation of bases, nucleosides, and oligonucleotides : Retention manipulation via micellar solutions and metal additives.
[49] A.F.Lecoq, C.Leuratti, E.Marafante, S.Di Biase, Proceedings of the 13th Int.Symp. on Capillary Chromatography, Riva del Garda, Italy (1991), 1206.
Analysis of nucleic acid derivatives by micellar electrokinetic chromatography.
[50] J.Tehrani, D.Davison, L.Friling, Poster presented at the 1st Int.Symp. on High Performance Capillary Electrophoresis, Boston, MS, USA (1989), T-P-12.
A high-performance variable UV-VIS Absorbance Detector for Capillary Electrophoresis.
[51] J.A.Lux, H.Yin, G.Schomburg, J.High Resolut.Chromatogr., **13** (1990) 145.
Influence of polymer coating of capillary surfaces on migration behavior in micellar electrokinetic capillary chromatography.
[52] K.H.Row, W.H.Griest, M.P.Maskarinec, J.Chromatogr., **409** (1987) 193.
Separation of modified nucleic acid constituents by micellar electrokinetic capillary chromatography.
[53] W.H.Griest, M.P.Maskarinec, K.H.Row, Sep.Sci.Technol., **23** (1988) 1905.
Evaluation of micellar electrokinetic capillary chromatography for the separation and detection of normal and modified deoxyribonucleosides and deoxyribonucleotides.
[54] T.Tsuda, K.Takagi, T.Watanabe, T.Satake, J.High Resolut.Chromatogr. Chromatogr.Commun., **11** (1988) 721.
Separation of nucleotides in organs of Guinea pig by capillary zone elctrophoresis.
[55] V.Dolnik, J.Liu, F.Banks, M.V.Novotny, P.Boček, J.Chromatogr., **480** (1989) 321.
Capillary zone electrophoresis of oligonucleotides. Factors affecting separations.
[56] L.Gross, E.S.Yeung, J.Chromatogr., **480** (1989) 169.
Indirect fluorimetric detection and quantification in capillary zone electrophoresis of inorganic anions and nucleotides.
[57] J.Liu, F.Banks, M.Novotny, J.Microcol.Sep., **1** (1989) 136.
High-speed micellar electrokinetic capillary chromatography of the common phosphorylated nucleosides.
[58] J.I.Ohms, Biotechnology Application Data DS-738, Beckman Instruments Inc., 1990.
Analysis of nucleic acid derivatives and analogs by P/ACE™ System 2000 using capillary electrophoresis.
[59] T.Lee, E.S.Yeung, M.Sharma, J.Chromatogr., **565** (1991) 197.
Micellar electrokinetic capillary chromatographic separation and laser-induced fluorescence detection of 2'-deoxynucleoside 5'-monophosphates of normal and modified bases.
[60] H.-F.Yin, J.A.Lux, G.Schomburg, J.High Resolut.Chromatogr., **13** (1990) 624.
Production of polyacrylamide gel filled capillaries for capillary gel electrophoresis (CGE) : Influence of capillary surface pretreatment on performance and stability.
[61] S.Fujiwara, S.Iwase, S.Honda, J.Chromatogr., **447** (1988) 133.
Analysis of water-soluble vitamins by micellar electrokinetic capillary chromatography.
[62] H.Nishi, N.Tsumagari, T.Kakimoto, S.Terabe, J.Chromatogr., **465** (1989) 331.
Separation of water-soluble vitamins by micellar electrokinetic chromatography.
[63] D.E.Burton, M.J.Sepaniak, M.P.Maskarinec, J.Chromatogr.Sci., **24** (1986) 347.
Analysis of B6 vitamers by micellar electrokinetic capillary chromatography with laser-excited fluorescence detection.
[64] S.Kobayashi, T.Ueda, M.Kikumoto, J.Chromatogr., **480** (1989) 179.
Photodiode array detection in high-performance capillary electrophoresis.
[65] H.Nishi, N.Tsumagari, S.Terabe, Anal.Chem., **61** (1989) 2434.
Effect of tetraalkylammonium salts on micellar electrokinetic chromatography of ionic substances.
[66] C.P.Ong, C.L.Ng, H.K.Lee, S.F.Y.Li, J.Chromatogr., **547** (1991) 419.
Separation of water- and fat-soluble vitamins by micellar electrokinetic chromatography.
[67] M.E.Swartz, J.Liq.Chromatogr., **14** (1991) 923.
Method development and selectivity control for small molecule pharmaceutical separations by capillary electrophoresis.

[68] W.Thormann, P.Meier, C.Marcolli, F.Binder, J.Chromatogr., **545** (1991) 445.
Analysis of barbiturates in human serum and urine by high-performance capillary electrophoresis-micellar electrokinetic capillary chromatography with on-column multi-wavelength detection.
[69] A.M.Hoyt, M.J.Sepaniak, Anal.Lett., **22** (1989) 861.
Determination of benzylpenicillin in pharmaceuticals by capillary zone electrophoresis.
[70] H.Nishi, N.Tsumagari, T.Kakimoto, S.Terabe, J.Chromatogr., **477** (1989) 259.
Separation of β-lactam antibiotics by micellar electrokinetic chromatography.
[71] H.Nishi, T.Fukuyama, M.Matsuo, J.Chromatogr., **515** (1990) 245.
Separation and determination of aspoxicillin in human plasma by micellar electrokinetic chromatography with direct sample injection.
[72] T.Nakagawa, T.Oda, A.Shibukawa, H.Tanaka, Chem.Pharm.Bull., **36** (1988) 1622.
Separation and determination of cefpiramide in human plasma by electrokinetic chromatography with a micellar solution and an open tubular fused silica capillary.
[73] T.Nakagawa, Y.Oda, A.Shibukawa, H.Fukuda, H.Tanaka, Chem.Pharm. Bull., **37** (1989) 707.
Electrokinetic chromatography for drug analysis, separation and determination of Cefpiramide in human plasma.
[74] M.Miyake, A.Shibukawa, T.Nakagawa, J.High Resolut.Chromatogr., **14** (1991) 181.
Simultaneous determination of creatinine and uric acid in human plasma and urine by micellar electrokinetic chromatography.
[75] H.Nishi, T.Fukuyama, M.Matsuo, S.Terabe, J.Chromatogr., **513** (1990) 279.
Separation and determination of lipophilic corticosteroids and benzothiazepin analogues by micellar electrokinetic chromatography using bile salts.
[76] H.Nishi, M.Matsuo, J.Liq.Chromatogr., **14** (1991) 973.
Separation of corticosteroids and aromatic hydrocarbons by cyclodextrin-modified micellar electrokinetic chromatography.
[77] D.Y.Pharr, Z.S.Fu, T.K.Smith, W.L.Hinze, Anal.Chem., **61** (1989) 275.
Solubilization of cyclodextrins for analytical applications.
[78] J.Vindevogel, P.Sandra, Anal.Chem., **63** (1991) 1530.
Resolution optimization in micellar electrokinetic chromatography : Use of Plackett-Burman statistical design for the analysis of testosterone esters.
[79] S.Fujiwara, S.Honda, Anal.Chem., **59** (1987) 2773.
Determination of ingredients of antipyretic analgesic preparations by micellar electrokinetic capillary chromatography.
[80] H.Nishi, T.Fukuyama, M.Matsuo, S.Terabe, J.Chromatogr., **498** (1990) 313.
Separation and determination of the ingredients of a cold medicine by micellar electrokinetic chromatography with bile salts.
[81] H.Nishi, T.Fukuyama, M.Matsuo, S.Terabe, J.Pharm.Sci., **72** (1990) 519.
Effect of surfactant structures on the separation of cold medicine ingredients by micellar electrokinetic chromatography.
[82] R.Weinberger, M.Albin, J.Liq.Chromatogr., **14** (1991) 953.
Quantitative micellar electrokinetic capillary chromatography : Linear dynamic range.
[83] R.A.Wallingford, A.G.Ewing, Anal.Chem., **60** (1988) 258.
Amperometric detection of catechols in capillary zone electrophoresis with normal and micellar solutions.
[84] R.A.Wallingford, A.G.Ewing, J.Chromatogr., **441** (1988) 299.
Retention of ionic and non-ionic catechols in capillary zone electrophoresis with micellar solutions.
[85] R.A.Wallingford, A.G.Ewing, Anal.Chem., **61** (1989) 98.
Separation of serotonin from catechols by capillary zone electrophoresis with electrochemical detection.
[86] R.A.Wallingford, P.D.Curry, A.G.Ewing, J.Microcol.Sep., **1** (1989) 23.
Retention of catechols in capillary electrophoresis with micellar and mixed micellar solutions.
[87] R.Weinberger, I.S.Lurie, Anal.Chem., **63** (1991) 823.
Micellar electrokinetic capillary chromatography of illicit drug substances.
[88] R.Weinberger, E.Sapp, S.Moring, J.Chromatogr., **516** (1990) 271.
Capillary electrophoresis of urinary porphyrins with absorption and fluorescence detection.
[89] I.Z.Atamna, G.M.Janini, G.M.Muschik, H.J.Issaq, J.Liq.Chromatogr., **14** (1991) 427.
Separations of xanthines and uric acids by capillary zone electrophoresis and micellar electrokinetic capillary electrophoresis.
[90] P.G.Pietta, P.L.Mauri, A.Rava, G.Sabbatini, J.Chromatogr., **549** (1991) 367.
Application of micellar electrokinetic chromatography to the determination of flavonoid drugs.

[91] J.Vindevogel, P.Sandra, L.C.Verhagen, J.High Resolut.Chromatogr., **13** (1990) 295.
Separation of hop bitter acids by capillary zone electrophoresis and micellar electrokinetic chromatography with UV-diode array detection.

[92] J.Vindevogel, P.Sandra, J.High Resolut.Chromatogr., **14** (1991) 795.
Micellar electrokinetic chromatography of ionic compounds : The analysis of hop bitter acids.

[93] J.Vindevogel, R.Szücs, P.Sandra, L.C.Verhagen, J.High Resolut. Chromatogr., **14** (1991) 584.
Analysis of beer iso-α-acids by micellar electrokinetic chromatography and multi-wavelength UV detection.

[94] J.Vindevogel, R.Szücs, P.Sandra, D.De Keukeleire, L.C.Verhagen, submitted,1991.
Micellar electrokinetic chromatography of hop bitter compounds.

[95] A.T.Balchunas, D.F.Swaile, A.C.Powell, M.J.Sepaniak, Sep.Sci.Technol., **23** (1988) 1891.
Separations of compounds of biological and environmental interest by micellar electrokinetic chromatography.

[96] S.Terabe, K.Otsuka, T.Ando, Anal.Chem., **57** (1985) 834.
Electrokinetic chromatography with micellar solution and open-tubular capillary.

[97] K.Otsuka, S.Terabe, T.Ando, J.Chromatogr., **348** (1985) 39.
Electrokinetic chromatography with micellar solutions. Retention behaviour and separation of chlorinated phenols.

[98] S.Terabe, K.Otsuka, K.Ichikawa, A.Tsuchiya, T.Ando, Anal.Chem., **56** (1984) 111.
Electrokinetic separations with micellar solutions and open-tubular capillaries.

[99] C.P.Ong, L.C.Ng, H.K.Chong, H.K.Lee, S.F.Y.Li, J.Chromatogr., **516** (1990) 263.
Retention of eleven priority phenols using micellar electrokinetic chromatography.

[100] C.D.Gaitonde, P.V.Pathak, J.Chromatogr., **514** (1990) 389.
Capillary zone electrophoretic separation of chlorophenols in industrial waste with on-column electrochemical detection.

[101] H.T.Rasmussen, H.M.McNair, J.High Resolut.Chromatogr. Chromatogr. Commun., **12** (1989) 635.
Influence of pH on elution behavior in micellar electrokinetic capillary chromatography.

[102] D.M.Northrop, D.E.Martire, W.A.MacCrehan, Anal.Chem., **63** (1991) 1038.
Separation and identification of organic gunshot and explosive constituents by micellar electrokinetic capillary electrophoresis.

[103] H.T.Rasmussen, L.K.Goebel, H.M.McNair, J.Chromatogr., **517** (1990) 549.
Micellar electrokinetic chromatography employing sodium alkyl sulfates and Brij 35.

[104] H.T.Rasmussen, L.K.Goebel, H.M.McNair, J.High Resolut.Chromatogr., **14** (1991) 25.
Optimization of resolution in micellar electrokinetic chromatography.

[105] F.R.Callejas, J.Vindevogel, P.Sandra, unpublished results.

[106] S.Terabe, Y.Miyashita, O.Shibata, E.R.Barnhart, L.R.Alexander, D.G.Patterson, B.L.Karger, K.Hosoya, N.Tanaka, J.Chromatogr., **516** (1990) 23.
Separation of highly hydrophobic compounds by cyclodextrin-modified micellar electrokinetic chromatography.

[107] M.W.F.Nielen, M.J.A.Mensink, J.High Resolut.Chromatogr., **14** (1991) 417.
Separation of benzothiazole sulfenamides using micellar electrokinetic capillary chromatography.

[108] T.Saitoh, H.Hoshino, T.Yotsuyanagi, J.Chromatogr., **469** (1989) 175.
Separation of 4-(2-pyridylazo)-resorcinolato metal chelates by micellar electrokinetic capillary chromatography.

[109] K.Saitoh, C.Kiyohara, N.Suzuki, J.High Resolut.Chromatogr., **14** (1991) 245.
Mobilities of metal-β-diketonate complexes in micellar electrokinetic chromatography.

13 Chiral Separations

13.1 Chiral Separations in Electrophoretic Techniques

Enantiomers behave identically in a non-chiral environment and their separation requires the introduction of a chiral selector . The same strategies as have been exploited in enantiomer resolution in liquid chromatography [1] can, and have been, used in electrokinetic analysis[1].

- *Formation of diastereomers.* Differentiation is obtained by preliminary reaction with a chiral derivatizing reagent. The resulting diastereomers are then separated in a conventional, achiral environment.
- *Chiral mobile phase additives.* Chiral selectivity is obtained by formation of transient diastereomeric complexes in the mobile phase. Selectivity can result from differences in equilibrium or from differences in retention of the complexes.
- *Direct enantiomer separation.* Chiral separation results from stereospecific interaction between the analyte and a chiral stationary phase.

The difference between the latter two may be less obvious in electrokinetic separations. Following the formalism of the pseudostationary phase, chiral micellar phases will be considered as chiral (pseudo-)phases. Cyclodextrins again are difficult to categorize. In the presence of a micellar phase, the non-micellized cyclodextrins can be considered as chiral mobile phase additives. In the absence of surfactants, they are a borderline case between a chiral solute and a chiral microphase.

13.2 The Separation of Diastereomers

In conventional chromatography, the derivatization effort and the cost of the chiral reagent, which must be of high enantiomeric purity, is justified by the fact that the need for relatively expensive chiral phases or additives is avoided. In miniaturized techniques like CZE and MEKC however, the cost of the chiral pseudophase is much reduced along with the required volume and also by the fact that it need not be bound to a support. This may explain why this technique is less

---

[1] *An additional mechanism that has been observed to lead to partial chiral separation in LC results from the chirality of the sample itself but is not further considered here [2].*

explored. It has been applied to the enantiomeric resolution of amino acids (**Figure 13.1**) after derivatization with Marfey's reagent (1-fluoro-2,4-dinitrophenyl-5-L-alanine amide, **Figure 13.2a**) [3]. Amino acid derivatives with a carboxylate side group, like those of Asp and Glu, could not be resolved in normal elution mode. Their separation was realized in reversed direction mode with a pH 3.3 buffer. The same derivatization procedure was also applied to di- and tripeptides.

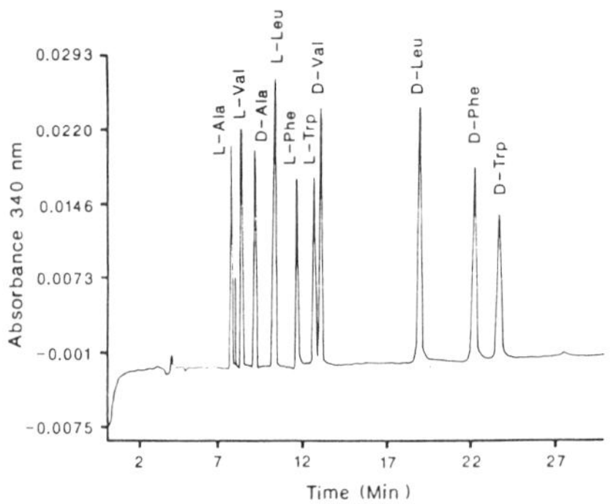

Figure 13.1
MEKC separation of racemic amino acids after derivatization with L-Marfey's reagent. Capillary : 75 µm x 50/57 cm. Buffer : 200 mM SDS in 100 mM borate (pH 8.5), containing 5% acetonitrile. Applied voltage : 20 kV. Detection : UV, 340 nm. *(Reprinted with permission [3] © Elsevier Science Publishers B.V.)*

Nishi et al. derivatized amino acids with 2,3,4,6-tetra-O-acetyl-β-D-glucopyranosyl isothiocyanate (GITC, **Figure 13.2b**) and separated the resulting diastereomeric thiourea derivatives with SDS [4]. Out of 21 amino acids, only Glu and Asp could not be resolved. For simultaneous separation, relatively hydrophilic amino acids are most difficult, which is ascribed to poor micellar solubilization. Nevertheless, simultaneous resolution and separation of 10 amino acid pairs has been realized (**Figure 13.3**).

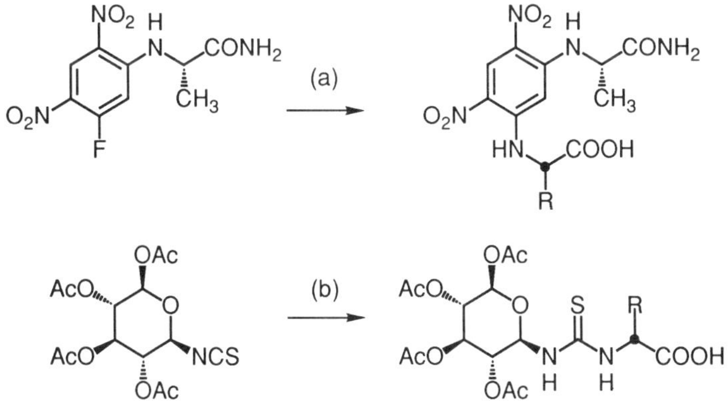

Figure 13.2
Formation of diastereomers for the chiral resolution of amino acids. (a) L-Marfey's reagent, (b) GITC.

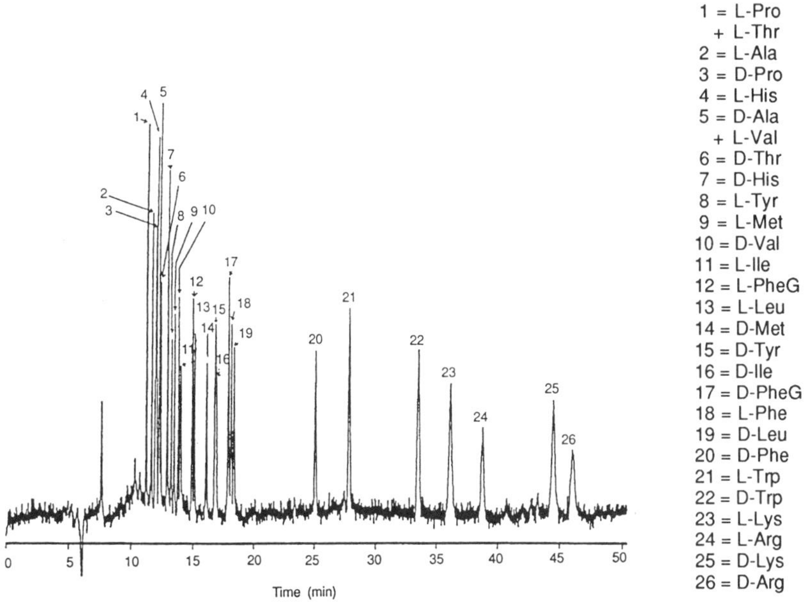

Figure 13.3
MEKC separation of 14 GITC-derivatized DL-amino acids. Capillary : 50 µm x 50/65 cm. Buffer : 250 mM SDS in 20 mM phosphate/borate (pH 9.0). Applied voltage : 20 kV. Detection : UV, 210 nm. *(Reprinted with permission [4] © MicroSeparations Inc.)*

13.3 Chiral Additives to the Aqueous Phase

Chiral separations in CZE were first demonstrated with dansylated amino acids in a Cu(II)-L-histidine containing buffer [5]. The selectivity arises from different equilibria in the ligand exchange between histidine and the D- and L-amino acids and the fact that the complexed amino acids show increased mobility. The separation of nine amino acid pairs was realized, although simultaneous analysis would have lead to some coelutions (like D-Met and L-α-aminobutyric acid). In subsequent investigations [6], the L-histidine was replaced by aspartame (L-aspartyl-L-phenylalanine methyl ester) (**Figure 13.4**). Of the 18 amino acid pairs tested, 14 were resolved and the elution time differences were larger than with the L-histidine complex. To increase the number of amino acids that can be resolved simultaneously, sodium tetradecyl sulfate was added to the buffer (**Figure 13.5**).

Like the copper-complexes, cyclodextrins can be used on their own for genuine electrophoretic separations. The first chiral separation in isotachophoresis was realized with cyclodextrins [7] and has been applied to drugs like pseudoephedrines [7-9], ketotifen [9,10], phenothiazins [8,9,11], benzothiepines [8] and chloramphenicol [9]. In these studies, the differences in selectivity between normal and methylated cyclodextrins have been described. Recently, the possibility of chiral separation of dansylated amino acids with cyclodextrins in transient-state isoelectric focusing has been reported [12].

Guttman et al. did not observe chiral selectivity for dansylated amino acids in CZE, when cyclodextrins were added to the buffer [13]. Reasoning that the differences between complexed and uncomplexed forms should be as large as possible, they realized effective zero mobility for the complexes by trapping the cyclodextrins in a polyacrylamide gel and resolved several dansylated amino acids. Since then, chiral CZE separations with free cyclodextrins have proved to be possible for ephedrine-related compounds [14], catecholamines [14,15], tryptophan [15], terbutaline [16] and propranolol [16]. Successful separation critically relies on size and degree of methylation of the cyclodextrins and can be improved by adding organic modifiers like methanol or urea.

The addition of cyclodextrins to micellar buffers (SDS) results in a complex system in which both inclusion and micellar solubilization determine resolution. The cyclodextrins reduce capacity factors as well as provoking enantiomeric recognition [17]. In CD-MEKC the larger γ-cyclodextrin often gives the best results, even for compounds that perform best with β-cyclodextrins in other techniques. It has been hypothised that surfactant monomers are also included in the cyclodextrin cavity and

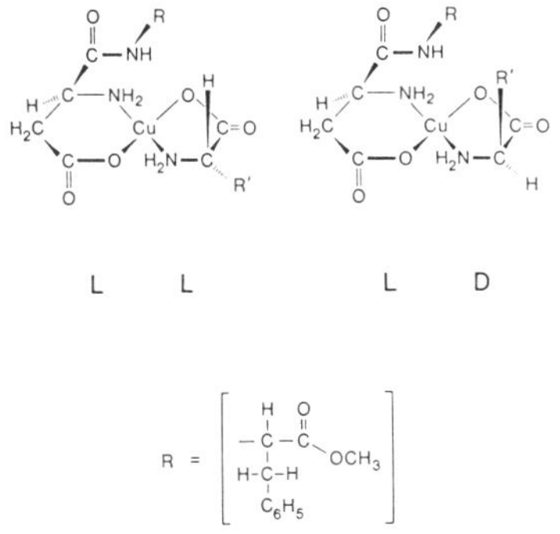

Figure 13.4
Cu(II) complexes formed with aspartame and dansylated amino acids. The D-amino acid complexes are generally more stable. *(Reprinted with permission [6] © American Chemical Society)*

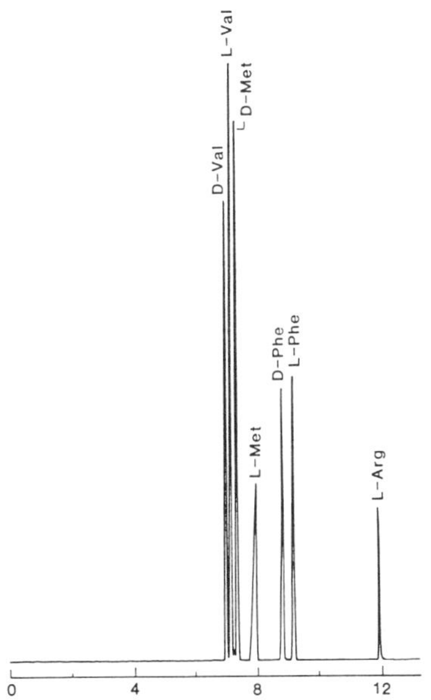

Figure 13.5
MEKC separation with laser-induced fluorescence detection (λ_{exc} = 325 nm) of dansylated amino acids. Capillary : 75 µm x 75/100 cm. Buffer : 20 mM STS in 2.5 mM $CuSO_4$, 5.0 mM aspartame and 10 mM NH_4OAc (pH 7.81). Applied voltage : 30 kV. *(Reprinted with permission [6] © American Chemical Society)*

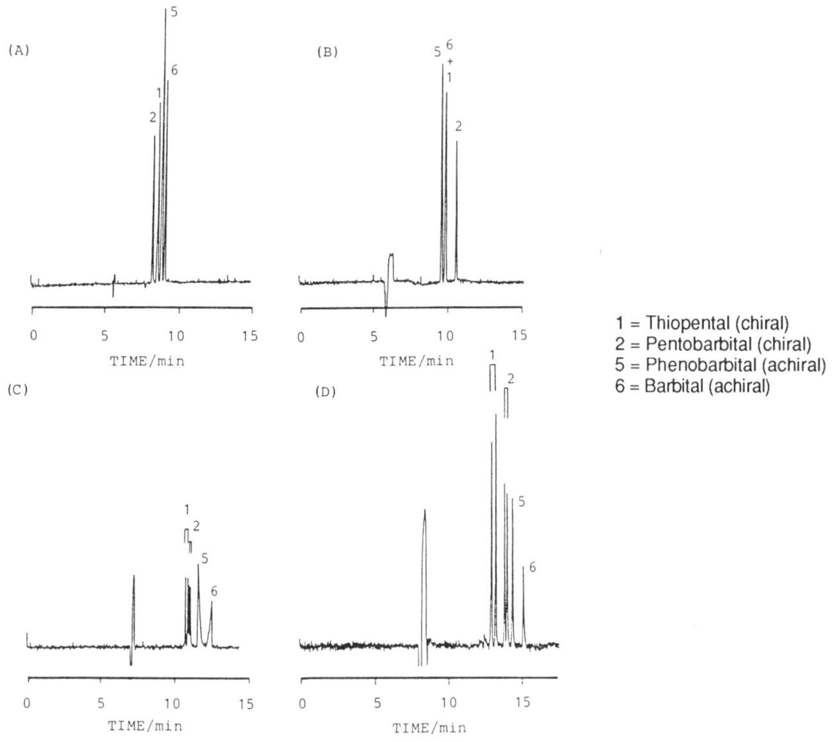

Figure 13.6
Separation of 4 barbiturates in different modes.
A = CZE, 20 mM phosphate/borate buffer, pH 9.0
B = MEKC, 50 mM SDS added to buffer in A
C = CD-MEKC, 30 mM γ-cyclodextrin added to buffer in B
D = CD-MEKC, 40 mM γ-cyclodextrin and 40 mM of d-camphor-10-sulfonate added to buffer in B.
Capillary : 50 μm x 50/65 cm. Applied voltage 20 kV. Detection : 220 nm. *(Reprinted with permission [17] © Elsevier Science Publishers B.V.)*

that a larger cavity is needed to include the solute together with the surfactant monomer [17]. For the few substrates that have been investigated until now (some baribiturates, binaphthyl compounds and trifluoro-1-(9-anthryl)ethanol), the addition of organic solvents like methanol reduced enantioselectivity although an improvement in peak shape was observed for the more hydrophobic solutes [17]. Chiral additives, however, like d-camphor-10-sulfonate and l-menthoxyacetic acid, improve enantioselectivity (**Figure 13.6**) which has been ascribed to simultaneous inclusion of both solute and additive in the γ-cyclodextrin cavity [17]. γ-CD-MEKC has also been applied to the resolution of dansylated amino acids [18,19].

13.4 Chiral Micellar Phases

13.4.1 Long-Hydrophobic Chain Chiral Surfactants

The first application of N-alkylated amino acids to form chiral micelles was briefly reported by *Cohen* et al. [20]. They mixed SDS with N,N-didecyl-L-alanine, in the presence of Cu(II) ions, and resolved several dansylated amino acids. A more detailed exploration of the possibility to use N-dodecanoyl-L-amino acid salts (pure or mixed with SDS), for chiral separations was reported by the group of *Hara* [21,22]. Of the two surfactants used, sodium dodecanoyl valinate (SDVal) and alaninate (SDAla), SDVal had the lowest CMC (3 vs. 5 mM) and showed also the highest enantioselectivity for the N-aroylamino acid isopropyl esters (**Figure 13.7**). Esterification of the substrates was essential for chiral resolution. Chiral selectivity increased with the number of nitrogroups in the aroyl moiety. Micelle concentration or the addition of methanol did not affect selectivity much and consequently,

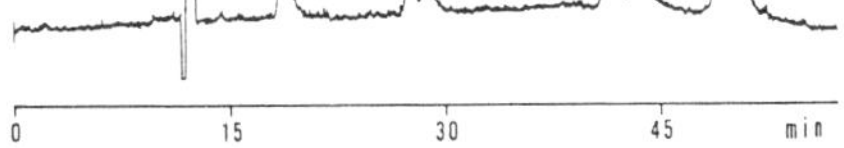

Figure 13.7
Chiral resolution of amino acids as their N-(3,5-dinitrobenzoyl)-O-isopropyl esters (D-enantiomers elute first). Capillary : 50 µm x 40/64 cm. Buffer : 12.5 mM SDVal and 12.5 mM SDS in phosphate/borate (pH 7.0), containing 10% of methanol. Applied voltage : 10.5 kV. Detection : UV, 254 nm. *(Reprinted with permission [21] © American Chemical Society)*

improved resolution must be attributed to better $\tilde{k}'$ regulation within the given elution range.

The efficiency in the SDVal system seems to be low and with phenylthiohydantoin amino acids, bad peak shapes and a lack of reproducibility have been observed [23,24]. On the addition of 10% methanol, peaks were still seriously tailed [24]. If besides 10% methanol, also 5 M of urea is added, peak tailing was further reduced [24]. The urea has is considered responsible for reducing wall adsorption [24], although no severe tailing problems were encountered in the achiral resolution of phenylthiohydantoin amino acids at a similar pH [25]. The chiral separation of benzoin and warfarin in the SDVal/methanol/urea system has also been reported [26].

13.4.2 Mixed Micelles with Nonionic Chiral Surfactants

In one case, a chiral pseudo-stationary phase has been obtained by mixing SDS with the nonionic surfactant digitonin, a glycoside of digitogenin. This reduces the electrophoretic mobility of the micelle so much that the elution range at neutral pH becomes small and no resolution is obtained. At pH 3.0, however, phenylthiohydantoin amino acids could be separated [23]. The analysis time for less solubilized derivatives tends to be long (**Figure 13.8**).

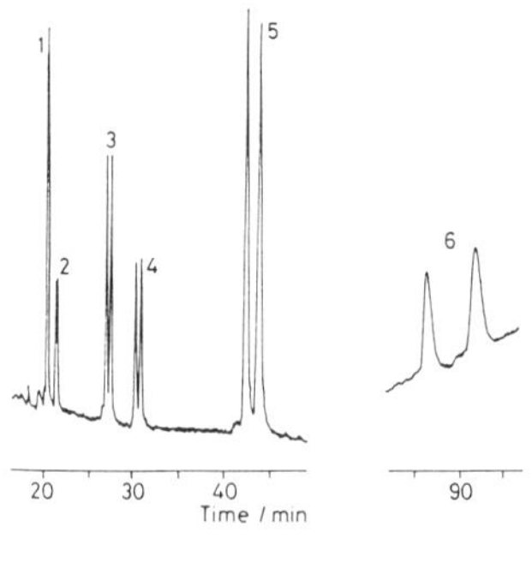

1 = PTH-tryptophan
2 = PTH-norleucine
3 = PTH-norvaline
4 = PTH-valine
5 = PTH-α-aminobutyric acid
6 = PTH-alanine

Figure 13.8

Chiral separation of PTH-DL-amino acids. Capillary : 50 μm x 49/63 cm. Buffer : 25 mM digitonin and 50 mM SDS in phosphate (pH 3.0). Applied voltage : -20 kV. Detection : UV, 260 nm. *(Reprinted with permission [23] © Elsevier Science Publishers B.V.)*

13.4.3 Bile Salts

The first reported chiral separations with bile salts required acidic pH and long analysis times. Several dansylated amino acids were resolved with taurocholate and taurodeoxycholate salts at pH 3.0 [27]. Analysis times, typically 60 to 80 minutes for more hydrophilic species as the norvaline and methionine derivatives, could be reduced by the addition of SDS but separation quality is not comparable to the one obtained in CD-MEKC (**Figure 13.9**).

Better results were obtained with several drugs (trimetoquinol, diltiazem and carboline derivatives) and other compounds (1-naphthylethylamine and 2,2'-dihydroxy-1,1'-dinaphthyl) [28]. Unlike the dansylated amino acids, none of these compounds was negatively charged at neutral pH (**Figure 13.10**) which results in more effective micellization and subsequent resolution at neutral pH. The anionic naproxen [28] or 1,1'-binaphthyl dicarboxylic acid [29] on the other hand could not be resolved. As with the dansylated amino acids, analysis in acid medium can reduce the charge and increase chiral selectivity [29].

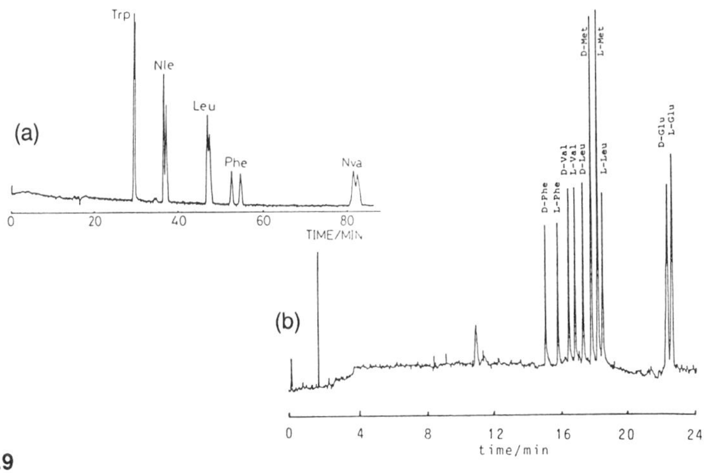

Figure 13.9
Chiral separation of dansyl DL-amino acids.
(a) 50 mM STDC in 50 mM phosphate, pH 3.0, -50 μA.
(b) 100 mM SDS and 60 mM γ-cyclodextrin in 100 mM borate, pH 8.3, +12 kV
Capillary : 50 μm x 50/70 cm (a) or 75 μm x 50/57 cm (b). Detection : UV, 210 (a) or 200 (b) nm. *(Reprinted with permission, a-[27] © Elsevier Science Publishers B.V., b-[30], Beckman Instruments Inc.)*

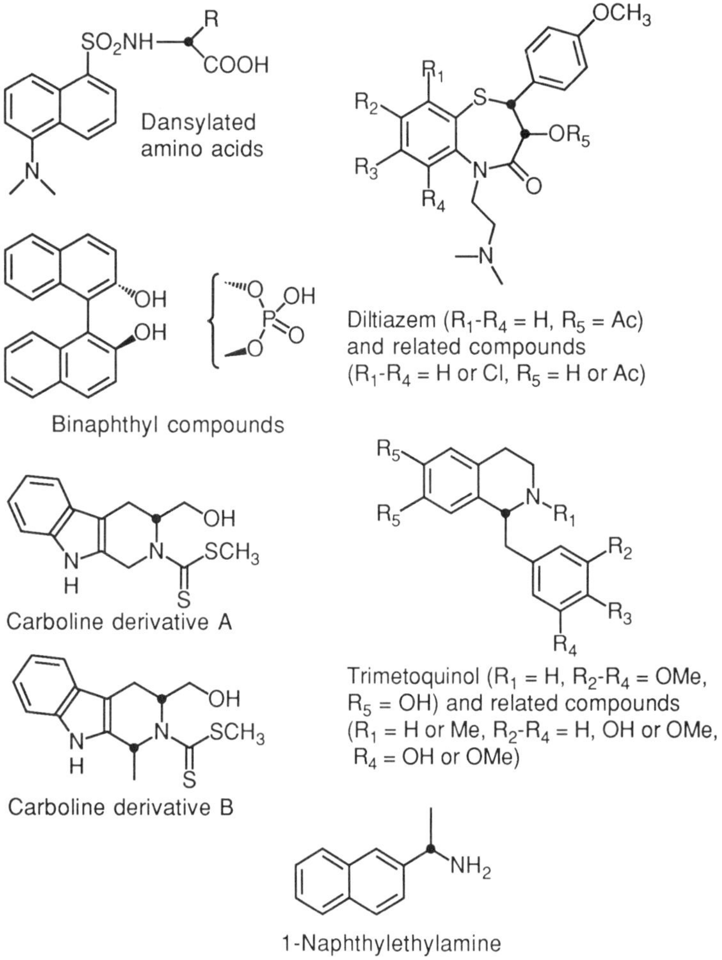

Figure 13.10
Chiral compounds resolved with bile salt micelles [27-29,31,32].

Simultaneous optimization of selectivity and capacity factor seems to be an important aspect of chiral separations with bile salts. The selectivity for carboline derivatives (**Figure 13.11**) did not vary much between (tauro-)cholates and (tauro-)deoxycholates but the much larger capacity factors in the latter case, beyond the optimal range, lead to unsuccessful resolution with SDC and STDC [28]. The deoxycholates (SDC and STDC) showed improved selectivity for the resolution of dihydroxydinaphthyl, when compared to SC and STC, and separation was more effective although the capacity factors were increased from 7-9 to 26-44 [28]. The influence of surfactant concentration and resolution optimization by adding methanol has been demonstrated [29]. Separation could not be obtained for 2,2,2-trifluoro-1-(9-anthryl)ethanol [28] but it is not clear whether this is due to lack of selectivity or to the total solubilization by the bile salts.

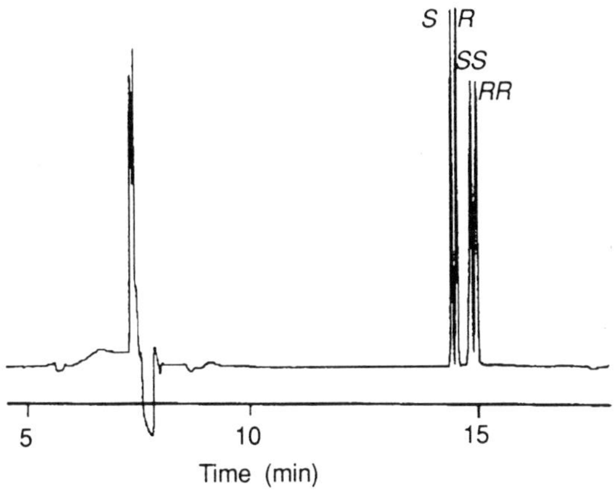

Figure 13.11
Chiral separation of R/S-Carboline derivative A and RR/SS-Carboline derivative B. Capillary : 50 µm x 50/65 cm. Buffer : 50 mM STC in 20mM phosphate/ borate (pH 9.0). Applied voltage : 20 kV. Detection : UV, 210 nm. *(Reprinted with permission [28] © MicroSeparations Inc.)*

The non-chiral determination of diltiazem with bile salts (SC) has already been mentioned (12.4.6). Only STDC seems capable of chiral recognition (**Figure 13.12**) [27,31] although partial resolution of D3 (R_1 = Cl, R_2-R_4 = H, R_5 = Ac, see Figure 13.10) has been observed with SC at pH 9.0 [31]. Trimetoquinol [27,31,32] and the related tetrahydropapaveroline [31] and laudanosine/norlaudanosoline (but not laudanosoline) [32] have been resolved with STDC at neutral pH. Other bile salts were not effective (see also 12.4.5) and the addition of urea, γ-cyclodextrin, glucose or sodium d-camphor-10-sulfonate to STDC solutions did not improve enantioselectivity [32]. MEKC with STDC has been applied to optical purity testing of S-trimetoquinol. R-impurities could be detected in the 1 to 3% range [31,32], a type of analysis which was not possible with LC on conventional chiral stationary phases [32].

The rational selection of a particular chiral pseudo-phase is not possible at this stage of development, but even in liquid chromatography, where much more experience has been accumulated, this is still not possible. It has been suggested that especially uncharged or basic solutes with a relatively rigid configuration would be suitable for enantiomeric separation with bile salts [27,31], but as already remarked, also capacity factor considerations must be taken into account.

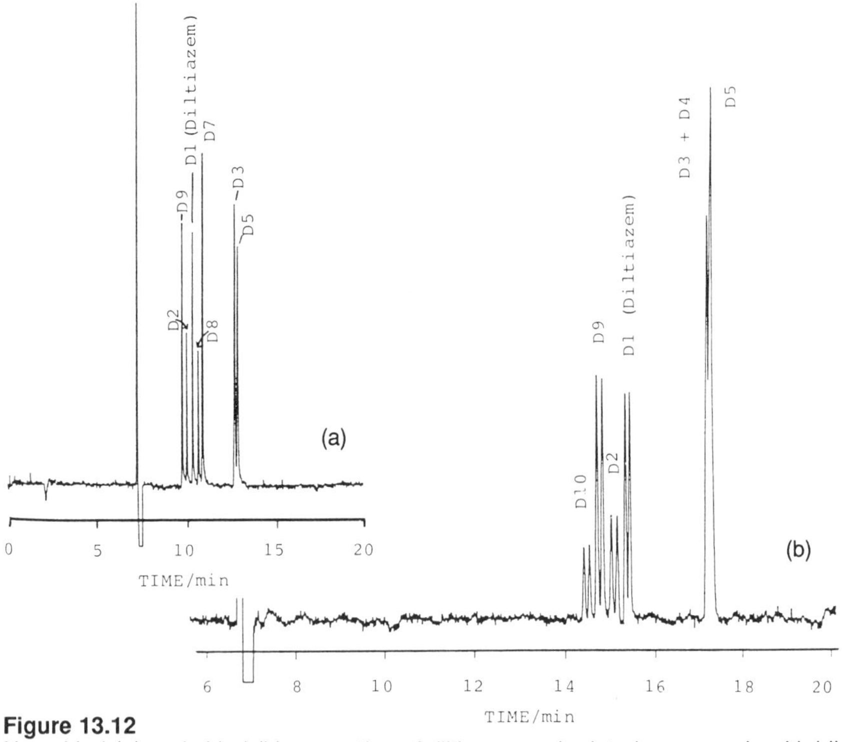

Figure 13.12
Non-chiral (a) and chiral (b) separation of diltiazem and related compounds with bile salts. Capillary : 50 µm x 50/65 cm. Applied voltage : 20 kV. Detection : UV, 210 nm. (a) 50 mM STC in 20 mM phosphate/borate, pH 7.0. (b) 50 mM STDC in 20 mM phosphate/borate, pH 7.0. *(Reprinted with permission [31] Elsevier Science Publishers B.V.)*

References

[1] R.W.Souter, "Chromatographic separations of stereoisomers", CRC Press Inc., Boca Raton, 1985
[2] A.Dobashi, Y.Motoyama, K.Kinoshita, S.Hara, Anal.Chem., **59** (1987) 2209.
Self-induced chiral recognition in the association of enantiomeric mixtures on silica gel chromatography.
[3] A.D.Tran, T.Blanc, E.Leopold, J.Chromatogr., **516** (1990) 241.
Free solution capillary electrophoresis and micellar electrokinetic resolution of amino acid enantiomers with L- and D-Marfey's reagent.
[4] H.Nishi, T.Fukuyama, M.Matsuo, J.Microcol.Sep., **2** (1990) 234.
Resolution of optical isomers of 2,3,4,6-tetra-O-acetyl-β-D-glucopyranosyl isothiocyanate (GITC)-derivatized DL-amino acids by micellar electrokinetic chromatography.
[5] E.Gassmann, J.E.Kuo, R.N.Zare, Science, **230** (1985) 813.
Electrokinetic separation of chiral compounds.
[6] P.Gozel, E.Gassmann, H.Michelsen, R.N.Zare, Anal.Chem., **59** (1987) 44.
Electrokinetic resolution of amino acid enantiomers with copper(II)-aspartame support electrolyte.
[7] J.Snopek, I.Jelínek, E.Smolková-Keulemansová, J.Chromatogr., **438** (1988) 211.
Use of cyclodextrins in isotachophoresis. IV. The influence of cyclodextrins on the chiral resolution of ephedrine alkaloid enantiomers.
[8] J.Snopek, I.Jelínek, E.Smolková-Keulemansová, J.Chromatogr., **472** (1989) 308.
Use of cyclodextrins in isotachophoresis. VIII. Two-dimensional chiral separations in isotachophoresis.
[9] J.Snopek, H.Soini, M.V.Novotny, Proceedings of the 13th Int.Symp. on Capillary Chromatography, Riva del Garda, Italy (1991), 1229.
Chiral resolution through cyclodextrin selectors : A comparison between capillary zone electrophoresis and isotachophoresis.
[10] I.Jelínek, J.Snopek, E.Smolková-Keulemansová, J.Chromatogr., **439** (1988) 386.
Use of cyclodextrins in isotachophoresis. V. The separation of ketotifen and its polar intermediate enantiomers.
[11] I.Jelínek, J.Dohnal, J.Snopek, E.Smolková-Keulemansová, J.Chromatogr., **464** (1989) 139.
Use of cyclodextrins in isotachophoresis. VII. Resolution of structurally related and chiral phenothiazines.
[12] P.G.Righetti, C.Ettori, P.Chafey, J.P.Wahrmann, Electrophoresis (Weinheim), **11** (1990) 1.
Enantiomer resolution in immobilized pH gradient gels via inclusion of a chiral discriminator.
[13] A.Guttman, A.Paulus, A.S.Cohen, N.Grinberg, B.L.Karger, J.Chromatogr., **448** (1988) 41.
Use of complexing agents for selective separation in high-performance capillary electrophoresis. Chiral resolution via cyclodextrins incorporated within polyacrylamide gel columns.
[14] S.Fanali, J.Chromatogr., **474** (1989) 441.
Separation of optical isomers by capillary zone electrophoresis based on host-guest complexation with cyclodextrines.
[15] S.Fanali, P.Boček, Electrophoresis (Weinheim), **11** (1990) 757.
Enantiomer resolution by using capillary zone electrophoresis : resolution of racemic tryptophan and determination of the enantiomer composition of commercial pharmaceutical epinephrine.
[16] S.Fanali, J.Chromatogr., **545** (1991) 437.
Use of cyclodextrins in capillary zone electrophoresis. Resolution of terbutaline and propranolol enantiomers.
[17] H.Nishi, T.Fukuyama, S.Terabe, J.Chromatogr., **553** (1991) 503.
Chiral separation by cyclodextrin modified micellar electrokinetic chromatography.
[18] S.Terabe, Y.Miyashita, Y.Ishima, K.Otsuka, Proceedings of the 13th Int.Symp. on Capillary Chromatography, Riva del Garda, Italy (1991), 962.
Enantiomeric separation by micellar electrokinetic chromatography.
[19] A.Trisciani, Proceedings of the 13th Int.Symp. on Capillary Chromatography, Riva del Garda, Italy (1991), 1703.
Optimization of cyclodextrin modified micellar electrokinetic capillary chromatography in the analysis of optically active dansyl amino acids.
[20] A.S.Cohen, A.Paulus, B.L.Karger, Chromatographia, **24** (1987) 15.
High-performance capillary electrophoresis using open tubes and gels.
[21] A.Dobashi, T.Ono, S.Hara, J.Yamaguchi, Anal.Chem., **61** (1989) 1984.
Optical resolution of enantiomers with chiral mixed micelles by electrokinetic chromatography.
[22] A.Dobashi, T.Ono, S.Hara, J.Yamaguchi, J.Chromatogr., **480** (1989) 413.

Enantioselective hydrophobic entanglement of enantiomeric solutes with chiral functionalized micelles by electrokinetic chromatography.
[23] K.Otsuka, S.Terabe, J.Chromatogr., **515** (1990) 221.
Enantiomeric resolution by micellar electrokinetic chromatography with chiral surfactants.
[24] K.Otsuka, S.Terabe, Electrophoresis (Weinheim), **11** (1990) 982.
Effects of methanol and urea on optical resolution of phenylthiohydantoin-DL-amino acids by micellar electrokinetic chromatography with sodium N-dodecanoyl-L-valinate.
[25] K.Otsuka, S.Terabe, T.Ando, J.Chromatogr., **332** (1985) 219.
Electrokinetic chromatography with micellar solutions. Separation of phenylthiohydantoin-amino acids.
[26] K.Otsuka, S.Terabe, Proceedings of the 12th Int.Symp. on Capillary Chromatography, Kobe, Japan (1990), 646.
Optical resolution by micellar electrokinetic chromatography with chiral surfactants.
[27] S.Terabe, M.Shibata, Y.Miyashita, J.Chromatogr., **480** (1989) 403.
Chiral separation by electrokinetic chromatography with bile salt micelles.
[28] H.Nishi, T.Fukuyama, M.Matsuo, S.Terabe, J.Microcol.Sep., **1** (1989) 234.
Chiral separation of optical isomeric drugs using micellar electrokinetic chromatography and bile salts.
[29] R.O.Cole, M.J.Sepaniak, W.L.Hinze, J.High Resolut.Chromatogr., **13** (1990) 579.
Optimization of binaphthyl enantiomer separations by capillary zone electrophoresis using mobile phases containing bile salts and organic solvent.
[30] Y.Miyashita, S.Terabe, Applications Data DS-767, Beckman Instruments Inc., 1990.
Separation of dansyl DL-amino acids by micellar electrokinetic capillary chromatography with and without cyclodextrins using P/ACE™ System 2000.
[31] H.Nishi, T.Fukuyama, M.Matsuo, S.Terabe, J.Chromatogr., **515** (1990) 233.
Chiral separation of diltiazem, trimetoquinol and related compounds by micellar electrokinetic chromatography with bile salts.
[32] H.Nishi, A.T.Fukuyama, M.Matsuo, S.Terabe, Anal.Chim.Acta, **236** (1990) 281.
Chiral separation of trimetoquinol hydrochloride and related compounds by micellar electrokinetic chromatography using sodium taurodeoxycholate solutions and application to optical purity determination.

Postscript

" ... and therefore Achilles can NEVER catch up with the Tortoise." (from a well-known Greek philosophical tale, paraphrased by D.R.Hofstadter)

As the chapters of this book were gradually finished, papers kept on appearing and by the time the first draft was ready, a 560-page symposium volume, containing 48 contributions on capillary electrophoretic techniques [1], became available. Incorporating the facts and findings of those papers in this book would have taken a considerable time and, moreover, during that time several new papers would undoubtedly have appeared. Any ancient Greek, who happens to be called Zeno, will prove that this race can never be won. Extrapolations from the 1981-1987 period lead to 1900 contributions on capillary electrophoretic techniques for the year 1992 [2]. As a wild guess, a few hundred papers of those 1900 (that is an average of several each week), will be related to MEKC, directly or at least in a general way. Under those circumstances, it would be a Sisyphean task to strive for completeness. On the other hand, the unrivalled growth prospect was considered a real possibility, taking into account the recent availability of commercial instrumentation [2]. For those who take this opportunity and now enter this new field in separation science, we hope to have provided the necessary background.

References

[1] J.W.Jorgenson (ed.), J.Chromatogr., **559** (1991) 1.
3rd International Symposium on High-Performance Capillary Electrophoresis, San Diego, CA, February 3-6, 1991.

[2] T.Braun, S.Nagydiósi-Rózsa, Trends Anal.Chem., **10** (1991) 266.
Capillary electrophoresis : Prospects for growth.

Appendix 1 : Simplified Model for Isotachophoresis

The following simplified isotachophoresis model [1] is based on the following principles :

- the current through adjacent zones must be equal
- all zones are moving with the same velocity
- in each zone, electroneutrality must be maintained.

First let us rewrite Ohm's law

$$I = \frac{V}{R} = \frac{V}{\frac{L}{\pi r^2}\rho} \tag{A1.1}$$

in terms of ionic mobilities. Because the diameter of the capillary is uniform, current can be replaced by current density ($i = I / \pi r^2$) :

$$i = \frac{V}{L}\kappa = E.F.\sum c_i |q_i| \mu_i \tag{A1.2}$$

The current density through the adjacent zones must be equal :

$$i_1 = i_2 = E_1.F.\sum c_{i,1} |q_{i,1}| \mu_{i,1} = E_2.F.\sum c_{i,2} |q_{i,2}| \mu_{i,2} \tag{A1.3}$$

or, considering only monovalent ions, and using unsigned values for the mobilities :

$$E_1.(c_{A,1}\,\mu_A + c_{Q,1}\,\mu_Q) = E_2.(c_{B,2}\,\mu_B + c_{Q,2}\,\mu_Q) \tag{A1.4}$$

To preserve electroneutrality in each zone :

$$c_{A,1} = c_{Q,1} \qquad \text{and} \qquad c_{B,2} = c_{Q,2} \tag{A1.5}$$

so that

$$E_1.c_{A,1}.(\mu_A + \mu_Q) = E_2.c_{B,2}.(\mu_B + \mu_Q) \tag{A1.6}$$

As the adjacent zones move at equal speed :

$$v_1 = E_1 . \mu_{A,1} = E_2 . \mu_{B,2} = v_2 \tag{A1.7}$$

so that finally :

$$c_{B,2} = c_{A,1} . \frac{\mu_B}{\mu_A} . \left(\frac{\mu_A + \mu_Q}{\mu_B + \mu_Q} \right) \tag{A1.8}$$

This is the Kohlrausch regulation function. More generally it can be formulated by the following equation which must be satisfied in each zone :

$$\sum_i \frac{c_i}{\mu_i} = \text{constant} \tag{A1.9}$$

It can be easily verified that, when expressed in terms of the example used,

$$\frac{c_{A,1}}{\mu_A} + \frac{c_{Q,1}}{\mu_Q} = \frac{c_{B,2}}{\mu_B} + \frac{c_{Q,2}}{\mu_Q} \; (= \text{constant}) \tag{A1.9}$$

using the electroneutrality requirement (eq.(A1.5)) leads to eq (A1.8).

References

[1] F.M.Everaerts, J.L.Beckers, Th.P.E.M.Verheggen, "Isotachophoresis : Theory, Instrumentation and Applications", Elsevier Science Publ., Amsterdam, 1976 (J.Chromatogr.Lib., vol.6).

Appendix 2 : List of Symbols and Abbreviations

A2.1 Symbols

Except for the most simple cases, SI units are specified to indicate dimensionality, but prefixed or alternative units are occasionaly used in the text. We have preferred the use of common symbols although this leads to some double uses (like α for selectivity and for the dissociation constant) which will, however, be self-explanatory through the context.

a	distance from the shear plane to the center of the ion	
aq	related to aqueous phase	
B_η	temperature dependence of viscosity ($\eta = A\ e^{B/T}$), for H_2O ca. 2400 K	
c, c_i	concentration	
d_c	column diameter	
d_{MC}	intermicelle distance	
D_m	diffusion coefficient	$m^2.sec^{-1}$
e	electronic charge	1.602×10^{-19} C
eo	electroosmotic	
ep	electrophoretic	
E	electric field	$V.m^{-1}$
F	Faraday's constant	96487 $C.mol^{-1}$
g	gravitational acceleration	9.81 $m.sec^{-2}$
g_f	geometric factor (4π for electroosmosis, 6π for electrophoresis)	
G_V	developed power per unit volume	$W.m^{-3}$
G_L	developed power per unit length	$W.m^{-1}$
h	plate height	
h_{inj}	height difference	
i	current density	$A.m^{-3}$
I	current	A
I	ionic strength	
inj	related to injection	
k	Boltzmann's constant	1.3805×10^{-23} $J.K^{-1}$
k', $\tilde{k}'$	capacity factor	
k_d	desorption rate	sec^{-1}
K_{eq}	distribution coefficient	
L	column length	

L_V	total column length (related to applied voltage)	
L_d	column length to detector (related to migration time)	
MC	related to micellar phase	
n, n_i	particle concentration	
n	peak capacity	
n_{aggr}	micelle aggregation number	
N	efficiency	
P	pressure	
q	ionic charge	
Q	quantity (weight)	
r	ionic radius	
r_c	capillary radius	
R	gas constant	$8.3143\ J.K^{-1}.mol^{-1}$
R_r	retention ratio	
R_s	resolution	
t_M	migration time (CZE)	
t_{MC}	micelle migration time	
t_R	retention time (MEKC)	
T	temperature	
v	linear velocity	
V	voltage	V
V_i	volume	
z, z_i	ion valency	
α	selectivity	
α	dissociation constant	
β	phase ratio	
δ^{-1}	Debye length	
ε	permittivity (= $4\pi\ \varepsilon_0.\varepsilon_r$)	$C^2.N^{-1}.m^{-2}$
ε_0	permittivity of vacuum	$8.85\ 10^{-12}\ C^2.N^{-1}.m^{-2}$
ε_r	relative permittivity or dielectric constant	± 78 for H_2O (25 C)
ζ	zeta potential	V
η	viscosity	$0.001\ kg.m^{-1}.sec^{-1}$ (H_2O, 25 C)
κ	conductivity	$S.m^{-1} = \Omega^{-1}.m^{-1}$
λ_T	thermal conductivity	$W.m^{-1}.K^{-1}$
Λ	equivalent conductivity	$m^2.\Omega^{-1}.mol^{-1}$
μ	mobility	$m^2.V^{-1}.sec^{-1}$

ρ	resistivity	Ω.m
σ	standard deviation	
σ_t	standard deviation (temporal)	
σ_L	standard deviation (spatial)	
φ	potential	V
ω	volume fraction	
ϖ	partial specific molar volume	$dm^3.mol^{-1}$

A2.2 Abbreviations

APS	Aminopropylsilane
BNA	3-Benzoyl-2-naphthaldehyde
CBQCA	3-(4-Carboxybenzoyl)-2-quinolinecarboxaldehyde
CD-MEKC	Cyclodextrin-modified Micellar Electrokinetic Chromatography
CEC	Capillary Electrochromatography
CGE	Capillary Gel Electrophoresis
CMC	Critical Micelle Concentration
CZE	Capillary Zone Electrophoresis
DNS	Dansyl
DABTH	Dimethylaminoazobenzene thiohydantoin
DNPH	N-2,4-Dinitrophenyl-hydrazone
DTAB	Dodecyltrimethylammonium Bromide
FITC	Fluorescein isothiocyanate
FTH	Fluorescein thiohydantoin
GITC	2,3,4,6-tetra-O-acetyl-β-D-glucopyranosyl isothiocyanate
HPLC	High-Perfcormance Liquid Chromatography
HTAB	Hexadecyltrimethylammonium Bromide
HQS	8-Hydroxyquinoline-5-sulfonic acid
ID	Internal diameter
ITP	Isotachophoresis
LC	Liquid Chromatography
MECC	Micellar Electrokinetic Capillary Chromatography
MEKC	Micellar Electrokinetic Chromatography
NBD	4-Chloro-7-nitrobenzo-2-oxa-1,3-diazole
NDA	2,3-Naphthalenedialdehyde
ODS	Octadecyltrichlorosilane, Octadecylated
OPA	Ortho-Phthaldialdehyde

PAGE	Polyacrylamide Gel Electrophoresis
PEG	Polyethylene glycol
PITC	Phenyl isothiocyanate
PTH	Phenylthiohydantoin
SC	Sodium Cholate
SDAla	Sodium Dodecanoyl Alaninate
SDC	Sodium Deoxycholate
SDHC	Sodium Dehydrocholate
SDS	Sodium Dodecyl Sulfate
SDVal	Sodium Dodecanoyl Valinate
STC	Sodium Taurocholate
STDC	Sodium Taurodeoxycholate
STDHC	Sodium Taurodehydrocholate
STLC	Sodium Taurolithocholate
STS	Sodium Tetradecyl Sulfate
TES	2-[Tris(hydroxymethyl)methylamino]-1-ethanesulfonic acid
TMCS	Trimethylchlorosilane
TRIS	Tris(hydroxymethyl)aminomethane

Acknowledgment

We thank the copyright owners for their kind permission to reproduce following figures and tables

- from *The Journal of MicroColumn Separations*, published by MicroSeparations, Inc. (Department of Chemistry, Brigham Young University, Provo, UT 84602, USA)

Figures 2.5, 5.10, 11.8, 13.3, 13.11

- from the *Journal of High Resolution Chromatography*, published by Hüthig Verlag (P.O.Box 102869, W-6900 Heidelberg, Germany)

Figures 2.6, 3.4, 5.6,5.7, 5.8,6.6, 6.9, 9.1, 9.2, 9.4
Table 5.2

- from *Analytical Chemistry*, published by The American Chemical Society (1155 Sixteenth St., N.W., Washington, DC 20036, USA)

Figures 3.3, 10.1, 10.3, 11.18, 11.19, 12.3, 12.10, 12.23, 13.4, 13.5, 13.7
Tables 11.2, 11.3

- from the *Journal of Chromatography*, published by Elseviers Science Publishers B.V. (Molenwerf 1, P.O.Box 211, 1000 AE Amsterdam, Nederland)

Figures 3.5, 3.6, 4.3, 4.4, 11.16, 11.20, 12.1, 12.5, 12.8, 12.9, 12.12, 12.13, 12.15, 12.16, 12.17a, 12.17b, 12.17d, 12.18, 12.20, 12.21, 12.22, 13.1, 13.6, 13.8, 13.9a, 13.12

- from the *Journal of Liquid Chromatography*, published by Marcel Dekker Inc. (270 Madison Avenue, New York, NY 10016, USA)

Figure 12.17c

- from *Beckman Applications Data* , published by Beckman Instruments Inc. (P.O.Box 10200, Palo Alto, CA 94304, USA)

Figures 12.4, 12.11, 13.9b

Subject Index

Compound derivatives are indicated by their abbreviations as listed in Appendix 2

Halvor Jaeger (Ed.)

Capillary Gas Chromatography - Mass Spectrometry in Medicine and Pharmacology

1987. XIII, 309 pp. 139 figs., 19 tabs.
Hardcover DM 108,—
ISBN 3-7785-1375-3

This book contains the proceedings of the recent Neu-Ulm conference on GC-MS in medicine and pharmacology. It therefore represents a significant addition to the literature of capillary gas chromatography and is intended as a practical aid and a reference work.

Contributions by leading scientists from many countries present the state of the art in GC-MS while providing sufficient background theory for a clear understanding of the practical applications. Analytical chemists in universities and industry and scientists working in the biomedical area will profit from this book.

From the contents:

W. G. Jennings and G. Takeoka: Recent Advances in Gas Chromatography; A. Sentissi, E. J. Rogers, R. Joppich-Kuhn, M. Joppich, and R. W. Giese: Derivatization of Iodothyronines at the Picogram Level for Determination by GC-ECD; W. A. Koenig: Application of Enantioselective Capillary Gas Chromatography to Stereochemical Analysis of Natural Compounds and Pharmaceuticals; H. M. Liebich: Organic Acids in Urine and Serum of Patients with Diabetic Ketoacidosis; W. Dünges: Some Remarks on Sample Preparation; D. Lutz and H. Jaeger: Applications of Capillary GLC as a Routine Analytical Technique in Biovailability Investigations; N. P. E. Vermeulen, W. Onkenhout, M. van der Graff, B. J. Xu, and A. G. L. Burm: Glass Capillary GLC-MS: Applications in the Analysis of Drugs and Metabolites in Biological Fluids; W. J. A. VandenHeuvel, R. W. Walker, and J. R. Carlin: Quantification of Biologically Active Compounds Using Stable Isotopes and Capillary Column GLC-Selected Ion Monitoring; N. Haering, J. A. Settlage, and R. Schubert: Measurement of Ergotamine in Human Plasma by Triple Stage Quadrupole MS with NICI; H. M. Leith and S. J. Gaskell: Highly Selective Detection of Trace Compounds During Capillary Gas Chromatography - Mass Spectrometry; C. Shackleton, S. J. Gaskell, and D. J. Liberato: Chromatographic (GC or LC) and Non-Chromatographic (FAB) Mass Spectrometric Techniques for Clinical Steroid Analysis; C. D. Pfaffenberger: Detection of Solventemia in Exposed Workers; H. Frank: Automation of Gas Chromatographic Analysis; R. E. Kaiser and R. I. Rieder: Results on Automatic Optimization of Gas Chromatography (Selectivity Optimization) by Closed Loop On-Line Computer Use; K. Jacob: Application of Capillary Gas Chromatography in Clinical Chemistry; K. Michaelis, J. A. Settlage, and H. Jaeger: Determination of Nitroglycerin and its Metabolites in Human Plasma.